9
R68.B

AF763419

FOOD HAZARDS AND FOOD HYGIENE

FOOD HAZARDS
AND
FOOD HYGIENE

Seema Yadav

ANMOL PUBLICATIONS PVT. LTD.
NEW DELHI - 110 002 (INDIA)

ANMOL PUBLICATIONS PVT. LTD.
H.O.: 4374/4B, Ansari Road, Darya Ganj,
New Delhi-110 002 (India)
Ph.: 23278000, 23261597
B.O.: No. 1015, Ist Main Road, BSK IIIrd Stage
IIIrd Phase, IIIrd Block,
Bangalore - 560 085 (India)
Visit us at: www.anmolpublications.com

Food Hazards and Food Hygiene

© 2006

[All rights reserved. No part of this publication may be reproduced, stored in a retrieval system or transmitted, in any form or by any means, mechanical, photocopying, recording or otherwise, without prior written permission of the publisher.]

PRINTED IN INDIA

Printed at Mehra Offset Press, Delhi.

Contents

Preface

The food we take is not always fit for human consumption. There is every chance that it contains certain harmful and undesirable substances, etc. It may be contaminated with various hazardous chemicals (insecticides/pesticides etc.) or may be contaminated with germs. A contaminated food if eaten may result in food poisoning. It is an extraordinary combination, food which nourishes us and poison which destroys us. Yet food poisoning is a hazard of every day life.

It is desirable that food poisoning never occur. The cause of food poisoning are germs and we know a lot about such germs. We must be able to destroy or control them.

In this book an attempt has been made to familiarise the general public about these facts and particularly the food handlers i.e. producers and distributors of food, manufacturers, of food products, shop keepers and cooks etc.

The life story of germs and their journey from farm to dining table has been discussed at length. An attempt has been made to bring home the important points and to caution the handler of food about the various dangers. The chapters on food hygiene brings about the importance of various hygienic conditions required in the food industry including restaurants and even the home kitchen.

It is expected that the book will serve the purpose of a layman, a textbook for undergraduate courses, a guide for teachers teaching in teachers training colleges. It can even serve as a science book for hygiene in school classes for the subjet of domestic science for girls.

Seema Yadav

1

Food Stuffs

1.1 *Introduction*

Animals need food mainly for (i) providing energy to the body for performing different life activities, (ii) building materials to maintain the body in living state, and (iii) to synthesize body substances required for growth and repair of tissues.

The major components of food are :

i) Carbohydrates	iv) Vitamins
ii) Fats	v) Minerals
iii) Proteins	vi) Water.

Food may be defined as any substance which when taken into the body has regulatory action on body.

1.2 *Food Intake*

Food as an energy supplier

All types of food can supply useful energy (i.e., the energy that is convertible into ATP or otherwise available for biological reactions). About 65% of energy combined in fats and carbohydrates can be trapped as ATP. The conversion factor is lower in case of proteins.

The *hunger centre* (H) and *satiety centre* (S) are the two eating control centres which regulate the food intake of animals. When hunger centre is stimulated it *incites* the body of the animal to eat and when satiety centre is stimulated it causes the animal to refuse food. The two centres function antagonistically to each other.

The *level of glucose* in the body is responsible to stimulate these eating control centres. High blood glucose level stimulates satiety centre and low blood glucose level stimulates the hunger centre.

The delivery of adequate amount of glucose to the body depends upon the following factors:

i) normal digestion
ii) normal liver function
iii) normal blood circulation
iv) normal hormone balance
v) normal nervous activities.

If any disturbance occurs in any body function it disturbs the food intake of the animal.

1.3 *Classification of Food*

Food can be classified as *diffusible* and *non-diffusible.*

Diffusible food includes minerals, salt, glucose, vitamins & water.

Non-diffusible food includes carbohydrates, lipids and proteins.

The food can also be grouped in another way as *energy food* and *non-energy food.*

Energy food includes proteins, lipids and carbohydrates.

Non-energy food includes vitamins, minerals, water and oxygen.

Food can also be classified as *inorganic food* and *organic food.*

Minerals are mainly *inorganic food* whereas *organic food includes* carbohydrates, lipids, proteins, fats etc.

1.4 *Functions of Food*

Eating is an essential part of the daily lives of everyone. Colleges and universities, recognizing the importance of food for the students, provide food service facilities on the campus. Even though the chief function of the food is to supply materials to meet physiological needs, it serves other functions also. Food plays an integral role in social life, whether on the college or university campus, in the homes of the land, or in the affairs of the convening diplomats of the world. Still another function of food is that of the psychological or emotional influence. Familiar foods are favourite foods, imparting a sense of well-being and satisfaction unrelated to food value but reminiscent of past experiences.

The Physiological Function of Food

The physiological function of food may be divided into three general categories: the need for food materials to supply energy, the need for food materials to build and maintain the cells and tissues, and the need for food materials to regulate body processes. These needs are satisfied by substances called *nutrients* which are found in the food man eats. The nutrients themselves have been subdivided into six general classes: carbohydrates, lipids, proteins, minerals, vitamins and water. The discussion of the part that these food factors play in the maintenance of good health will follow in subsequent chapters.

The Need for Food Nutrients to Supply Energy

Inasmuch as the primary function of food is to supply the body with energy-producing materials, the energy demand must be satisfied before the body uses food for building and maintenance, or regulation. The nutrients in foods which supply energy are called the *energy nutrients;* they include the carbohydrates, fats and proteins. Carbohydrate in the diet is the chief source of energy. This is true of diets around the world. Fat follows carbohydrate as a source of energy, with protein providing the least amount.

The Need for Food Nutrients to Build and Maintain Body Tissue

Nutrients used in the building and maintenance of the body are proteins, minerals and water. The most abundant body-building nutrient is water, followed by protein, with minerals occurring in the least amount. Each cell and tissue contain some of each of the body-building nutrients.

The Need for Food Nutrients to Regulate Body Processes

All of the nutrients, with the exception of carbohydrate, play a role in the regulation of the body processes. The essential fatty acids present in certain fats, the proteins, the minerals, the vitamins, and water each perform certain regulatory functions essential to normal body operation, such as the movement of fluids, the control of the balance between acid and base, the coagulation of blood, the activation of enzymes, and the maintenance of normal body temperature.

The Social Function of Food

Food has always served an important function in the social exchanges of people. In the ancient world guests could be invited only when there was enough food for current needs and future wants. The social occasions centering around food in our country are many – the sugaring off parties of Vermont and New Hampshire at maple sugar time, the "at homes" at Christmas and New Year's in the South, the food festivals of the Middle West and the outdoor hospitality of the Far West.

Inasmuch as food is an integral part of the social phase of college or university living, this function is important to college men and women.

Food is served at many social events: teas, breakfasts, banquets, athletic award dinners, dances and meetings of all sorts. At all of these affairs, food indirectly serves as an instrument to develop social rapport. The indirect purpose of refreshments after a club meeting may be to attract more students to the meeting, but the main function is to create an atmosphere where social relationships may be fostered. When a college man or coed is asked to plan a menu for a social function, the criterion for selection will probably *not* be whether the food will provide important nutrients needed for good nutrition but rather whether it will be popular with the guests.

1.5 Food Economics

In the years since that essay was reprinted, chemistry has accomplished more than could have been foreseen to guide the use of that most outstanding of all material resources, the food supply, in such manner as to make healthier and happier lives, more satisfying and of greater social value.

The spending of money for food, and the judgment of the proper place of food in the total budget of the individual or the family, have both been greatly enlightened by recent chemical investigations of nutritive values and of the relations of nutrition to health and to the extension of the prime of life.

With increased income, a part of the increase is used to buy more of the food which the people concerned most enjoy eating; but at any given level of income the accustomed expenditures for other things are too apt to be regarded as fixed by "the standard of living." And not only in the food trades but also among those who teach or investigate food science or food economics there often appears a sort of "vested interest" attitude of resentment against any teaching of readjustment of consumer's expenditure.

For such reasons it may be worthwhile to emphasize again the fact that we can use the guidance of the newer knowledge of nutrition and continue to eat every kind of food. As McCollum teaches: Eat what one should, and one may also eat what one likes; or, "Eat what you like so long as you eat what you should."

When the newer chemistry of nutrition suggests a greater prominence of fruits, of vegetables, and of milk (including cheese, cream and ice cream) in the American dietary, it is not suggesting anything untried.

Gradually increasing numbers of American families have already found satisfaction in this more modern plan, and in the superior health and efficiency which it includes. A gradual trend of the national food consumption in the direction indicated as beneficial by our newer nutritional knowledge means that the example of those who are already benefiting by this knowledge is being followed by one after another of their neighbours.

And the trend to greater prominence of fruits, vegetables, and dairy products in the family dietary and national food economy is progress for agriculture also. As expressed in an official publication of the U. S. Department of Agriculture, the general use of such dietaries "would not only improve the health and efficiency of the population, but at the same time would foster the type of agriculture which represents wise utilization of land for the country as a whole."

Partly but not entirely because of the recent evidence that the optimal calcium intake is significantly higher than is provided by hitherto approved dietary standards, increasing numbers of people guided by this newer chemistry of nutrition will doubtless follow the advice frequently given by McCollum and include a quart of milk per day in their customary food consumption.

Together with increased milk consumption there will probably continue to go an increased consumption of fruits and vegetables, especially the citrus fruits and the green vegetables: hitherto regarded as luxuries but now seen to be good dietary investments. Because fruits and vegetables differ so much among themselves and in the ways that they are served, it is harder here than in the case of milk to compare the consumption levels of different individuals or families in as accurate a way. Yet there need be no doubt that with growing acceptance of the guidance which our newer knowledge offers, fruits and vegetables are being given an ever greater prominence in the dietary.

Certainly many and probably most families and individuals could build health to higher levels by giving to fruit and to milk in its various forms a larger place than they now hold in the dietary. This is possible, even with present purchasing power, by moderate readjustment of expenditure: partly by shifts within the food budget; partly by economizing in other directions in order to have more to spend for food. Cost-of-living studies show that even among people of the lower income groups some money could be shifted to the purchase of more fruit and of more of whatever form of milk, cheese, cream, or ice cream one likes best, from present expenditures for adornment-other-than-clothing and amusement-other-than-automobile. Once the significance of the gains to health and efficiency is understood, this does not mean deprivation but rather increase of satisfaction. Fruits and dairy products can be so selected and used as to add much to the pleasures of the table. And the satisfaction of improving the buoyancy of health, the efficiency, and the level of achievement of all its members is the highest and most lasting pleasure that a family can derive from its income.

1.6 Significance of Home Gardens

Doubtless it is true that what is grown in home gardens will have no appreciable effect upon any important food industry. Recognition of this should pressure the agricultural economist and prevent undue expectations in the home gardener. The agricultural economist need not fear, and the home gardener will be wiser not to hope, that the products of home gardens will materially influence the food markets.

Yet the home gardens can be of very real and significant benefit through the fact that each family with a garden can grow an *extra* supply of fruits and vegetables to be consumed in addition to what they have been accustomed to feel able to buy. This can mean very material improvement of nutritional well-being and resultant health, happiness and efficiency; and the means for it can be found by a constantly increasing proportion of people as they come to realize how great a benefit it can be to themselves and (still more) to their children.

An important point here, which could not be clearly realized before we had the light of the newer chemistry of nutrition, is that, when the demand of the market has been fully met, in the agricultural-economic sense that consumers have bought all that they feel able to buy at a price at which producers and distributors can afford to sell, the average consumer family still does not have as much of fruit and fresh vegetables as it needs if it is to realize its full potentiality of health, efficiency and earning power.

1.7 The Fight for Food

Man the Hunter

All animals hunt, eat and sleep. Primitive man did just that — he caught his food, ate it, and slept it off. He added a few roots and berries and perhaps some leaves and nuts to his diet of flesh, and may be he learned to store some of these against a hungry period; but most days he was up and out and after his prey, and the quest for food was, as for all other animals, his one and only occupation. It is not known whether he ever suffered from food poisoning. If he did, the outbreak was limited to the occupants of his cave, for the cave-dweller hunted only for his own, and did not set up as a wholesaler or distributor of his kill. On the walls of his cave he drew the animals he hunted, and he may have hoped thereby to influence the luck of the chase; if so, it was his only attempt to manipulate his food supply, for this was otherwise a ruthless affair of hunting, killing and eating. There was neither time nor opportunity for sophistication.

Business Man

Modern man has no time to hunt. He is too busy checking invoices, cabling foreign brokers, or feeding his computers. Ships sail into his ports and, if there is no strike on, vast cargoes are unloaded into warehouses, and giant lorries pound their way thence down motorways to inland markets. At home he has long since regimented his animals and, instead of hunting them, he confines them within fences or buildings. He cares for them indeed, exhibits them at shows and festoons them with medallions. At the same time he grossly exploits their natural functions, especially those of procreation and milk production. When their time comes, he does not fell them where they graze, but ships them in trucks to markets where they are herded with truck-loads of other animals, also on their last pilgrimage. Finally, they reach the abattoir lairs, and finish up as so much dead meat on the slaughter-house floor. It is a highly complicated business, and much more ruthless than the primitive hunt.

Commercial Man

Modern man has also organized the products of the earth. He no longer accepts nature as he finds her, but rips and tears the land with steel, strews it with chemicals, and allows only those plants to grow which measure up to his demands. When the crops are ripe, they are not eaten by the man who grew them, but go off in lorry-loads to mills, markets, warehouses, distributors, food factories and retailers. It may be months or years before the final product reaches the table, where it may form an unrecognizable ingredient of some *plat du jour*, and be eaten by a dinner so civilized that he has forgotten that the delicacy before him comes from the field and the chase. All this is, of course, modern commerce, and without it modern life could not go on, but it brings problems. The caveman might occasionally come to grief from eating a poisonous berry, or a chunk of killed carcass left too long in his cave: but modern man puts a vast amount of manipulation and sophistication into the production of his food, and in the process runs far greater risks than his less demanding cave-man ancestor.

Man the Animal

Man is an animal. No matter how modern he may be, or how highly civilized may be the life he leads, he still needs food suitable for an animal. Unfortunately many other animals like the same kind of food. Rats flourish if they can get their teeth into food intended for humans, and a mouse will risk its life for a piece of cheese. Dogs and cats often share their owner's food, though cats can be fussier than their owners about what they will or will not eat; yet a tin of cat food differs very little, as far as proteins, fat or carbohydrates are concerned, from a tin of meat for humans. The main difference is in the label.

Germs Share his Food

The greatest menace to man's food supply comes, however, not from cats, dog, rats, mice and other visible animals, but from the hordes of invisible living organisms that also like the food man eats. These organisms are usually called micro-organisms, because they can be seen only under a microscope, but it is simpler to call them just germs. A germ consists of one cell only — it is unicellular. Man's body consists of countless numbers of cells — he is a multicellular animal. Yet a unicellular germ can cause disease or even death in multicellular man, if it can attack him under favourable conditions. Access to man's food supply provides such favourable conditions and all the complications of modern food production make such access all too easy. Some germs flourish and multiply enormously in food, and when an unsuspecting person eats some of it, he may go down with food poisoning, or even something worse like typhoid fever.

The germs that invade our food belong to many different families or races, and, just like humans, they have different habits and modes of life. Some germs called staphylococci find a cream bun or custard trifle much to their liking, provided it is neither too hot nor too cold. Others such as *Clostridium welchii* revel in a luke-warm stew. Salmonellae like meat, especially cold meat; if they can get inside a nice meat pie after it is cooked and cold, they begin to enjoy life, and multiply rapidly, but they hate heat. If the pie is well heated before it is eaten, the salmonellae are all killed and cause no harm, whereas if the pie is eaten cold, the unfortunate person who eats it may get salmonella food poisoning. Some germs produce poisonous substances called toxins while they are multiplying in food, and these toxins may be heat-resistant. When such food is heated the germs may be killed, but the toxins may not be destroyed, and the food, though it contains no live germs, may still cause food poisoning.

Clearly, before we can attack all these germs, we must know quite a lot about them. The general of an army spends a great deal of time studying his enemy; he tries to forecast how his enemy will behave, and then he plans his own maneuvers to try to upset his enemy's. In the same way, to prevent food poisoning we must study food-poisoning germs, trying to understand exactly how they get into our food, and how they behave when they are in. Only then can we adapt our methods of food production and preparation to prevent these germs from spoiling our food. This is what is involved in food hygiene, and this is what this book is about.

2

Food Habits and Their Significance

2.1 *The Eating Habits of College Students*

Certain general patterns of food selection characterize the diets of college students. Surveys show that as a rule certain nutrients are more generously supplied in the diets than others. Proteins and niacin are more generously supplied than any of the nutrients, reflecting an adequate intake of foods in the meat, fish, and poultry group; on the other hand, calcium, ascorbic acid, and vitamin A usually meet the Recommended Allowances the least well. Such shortcomings in the diet indicate that the students were eating less than the desirable amounts of milk and milk products, foods of the citrus fruit and tomato group, and leafy green and yellow vegetables. The thiamine intake is influenced to considerable extent by the caloric value of the diet; students whose diets contain less than the recommended energy intakes generally also have less than the recommended thiamine intakes.

Nutritionists at the University of Rhode Island wished to know why the ascorbic acid content of the diets of the coeds whom they were observing at that institution failed so frequently to meet the NRC recommendations—more often than any other nutrient except for iron. They found on investigation that the supply of ascorbic acid available to the students in the dormitory fare was adequate, if from the fruits provided for breakfast they chose a citrus rather than another fruit. Seldom did the intake of ascorbic acid meet the Recommended Allowance if citrus fruit was not eaten during the day.

College men eat better diets than college women in terms of meeting the Recommended Allowances. There is seldom a deviation from this pattern in all of the survey studies. Men drink more milk and eat more milk products and as a consequence, have a more satisfactory calcium and riboflavin intake than women. Ascorbic acid is less well supplied in the diets of women than of men. Most men have an acceptable intake of iron; many women do not. Diet records show as many as 36 to 42 per cent of the diets of college women provide less than two-thirds of the Recommended Allowance for iron. The unfortunate circumstance in this shortage of iron in the diets of young women is that this mineral is needed to take care of the continuing loss of iron through menstruation.

Skipping breakfast or eating an incomplete breakfast is reflected in the total nutrient intake for the day. Students at Montana State University who omitted breakfast or ate one that failed to meet the meal pattern of a fruit, a cereal, and eggs and/or meat, has a less adequate total nutrient content in the day's diet than those who did eat a complete breakfast. The missing nutrients as a result of no morning meal, or a poor one, were not made up in the later meals of the day.

2.2 *Food Consumption of Families*

In the spring of 1955 a nationwide survey of the food consumption of families in this country was made by the Household Economics Research Division and Market Development Branch of the U.S. Department of Agriculture. Families were interviewed by trained workers to obtain information on the food practices of the family. The survey was the most comprehensive to have been undertaken in this country up to that time.

The findings from the survey are representative of the national food habits at a time when there was a abundant supply of food and full employment. The markets offered a variety of products at various stages of preparation and at a wide range of prices. On the whole, the nutritive value of the diets was good. The adequacy of the food consumed reflected in some ways the general high level of purchasing power. Protein and niacin were most adequately provided, indicating a good supply of foods from the meat group of commodities. However, even with the abundance of food and the ability to buy the foods of choice in most cases, some nutrients were still inadequately supplied. Calcium and ascorbic acid were provided in less than the recommended amounts more frequently than all other constituents.

There were differences in the food consumption patterns and adequacy of the diets of families over the nation. These differences were related to place of dwelling (whether urban or farm), to family income, and to family size. How families eat is found to be related also to the home makers nutritional knowledge and level of education. Also sex and age of the individual make a difference in his food habits.

2.3 *Influence of Urban or Rural Living*

Farm families were found to have a higher energy intake than non-farm families, which may have been influenced by the family's outdoor physical labour and the abundant supply of home-produced foods. When the value of the home-produced foods was estimated at the price the family would pay for the same food on the market, it amounted on an average to around 40 per cent of the total cost of the food consumed. It should be noted that at the time of this survey, the farm family was producing less of their own food than in previous periods when surveys were made, reflecting a trend in the farm family food supply. During the depression of the thirties and the wartime of the forties, home gardens were imperative for food supply, in contrast to the abundance of food when the 1955 survey was made. Another such extensive survey is in progress now (1965). The chief differences between the urban and farm family diets in the kinds of foods consumed was the use of greater quantities of dairy products, grains, sugars, and fats by farm families and relatively less meat.

2.4 *Influence of Income*

Surveys of food consumption practices show a definite relationship between the income of the family and the amount expended for food. As the income goes up the amount of money spent for food increases but at a proportionately slower rate so that the per cent of the income going for food decreases. High-income families spend a lesser percent for food than those with lower incomes. Some of the increase in expenditure at higher income levels goes for larger quantities of foods; much of it goes for increased services in the form of foods requiring less preparation time by the home maker, as well as buying meals away from home. Families with more money also tend to choose more costly foods within a food group. For example, in the consumption study in 1955, families with an annual income between $2000 and $3000 paid, on the average, 56 cents a pound for meat; those with incomes from $4000 to $6000 paid 62 cents; and the families with a $6000 to $8000 annual income spent 66 cents. The cheaper cuts of meat were used by low income families; steak (other than round) appeared on the menu during the week in 24 per cent of the low-income families, and in 37 and 44 per cent, respectively, of the medium and high-income households.

The adequacy of the diet improves with increase in income. Vitamin A, riboflavin, and ascorbic acid, selected as representative of the diet as a whole, increased in adequacy as the income increased. It is true that spending more money for food tends to provide a better diet, but at each income level above subsistence in this extensive survey, some families were eating diets of good quality whereas others were choosing poor diets. Some of the families with incomes in the top third failed to meet the NRC recommendations. Most of the diets of these high-income families provided sufficient protein and niacin, but as many as 23 per cent of the diets failed to supply the recommended amount of calcium.

2.5 *Influence of Family Size*

The survey confirmed the already well-established relationship that large families spend more for food in total, but the amount expended for each family member is less than in small families. To illustrate, city families were selected in the income bracket of $4000 to $4999, with the number in the family increasing from two to six. The expenditure for food in terms of per cent of income increased, and the money value of the food per person decreased as the number in the family increased.

There are some explanations for the smaller expenditure per person in larger families. It is possible as a rule, to prepare food more economically for a large number of persons. Also in large families there are usually small children who require less of some of the foods. Furthermore, the quantities of foods, especially those that are higher in price, little difference in how well the diets of the family met the Recommended Allowances; however, more money was spent for food by the employed than non-employed home makers.

2.6 *Influence of Sex and Age*

Within the family group some members fare better than others in the adequacy of their diets. As a general rule boys and men consume relatively better diets than the girls and women. These data from a study in Groton Township, New York, illustrate the common pattern of food selection within a family group. As a whole, the young children (age groups under 10 years) fare best, followed next by the rest of the male members of the family. The least adequate diets were among the adolescent girls and women of childbearing age.

2.7 *Other Influences on Food Habits*

Family tradition and cultural background exert an influence on food habits which is less tangible than the influences discussed in the preceding paragraphs but real none the less. Another factor, and the ultimate determining one, is availability of food. Whatever it may be that prompts people to eat as they do, whether it be sheer availability of food, religion, the practices of their forebears, or even prejudices that may have developed, their established eating customs are an intimate part of them. The oneness of customs of eating with the person himself gives a feeling of security, a sense of well-being, and of "rightness in the world" when he is able to follow them; on the other hand, inability to get certain foods or to prepare them as desired may engender frustration.

The following account of an Afghanistan farmer could be told of most anyone in most any country. Improved maize was introduced that yielded approximately three times as much as the local varieties. Resistance to the new crop was strong; the reason—the taste of the bread compared unfavourably with the taste of bread made using traditional maize.

As an aid in menu planning, the Department of the Army studied the food preferences of the men. They found the best-liked food to be fresh milk, with hot rolls, hot biscuits, strawberry shortcake, and grilled steak close seconds. The least-like foods as a group were vegetables. What the investigators concluded about the eating habits of the army man is not reserved for him alone, but it would seem for all: "He eats what is available, what he likes, what his culture defines as food, what his personal history dictates, and what society and his peers say he should eat."

"What the culture defines as food" notably influences choice of food. Horseflesh, an edible substance, is such an example. In ancient times horsemeat was more widely used as a human food than it is today. The origin of today's prejudice is not known; some believe it might "derive from the animal's high status and its supposed holy qualities and association with deities". Another example is the case of the spotted pig; in the Chin States of Upper Burma, the small black local pig was crossed with a better meat producer. The offspring was spotted, and for that reason, by tradition unfit for eating.

Recognizing the fundamental nature of people's eating habits helps to understand their resistance to change. Dr. Aykroyd, speaking before the Third International Congress of Dietetics, clearly focused on the problem when he stated, "Each group tends to regard its food and meal patterns as the normal and natural ones." It may seem that refusing unfamiliar food by persons in need is reprehensible and unappreciative on their part, but the response is most likely deep seated and can only be changed by education and time.

The food habits most deeply entrenched are those followed by people living in isolated areas where the same food supply has prevailed for many years. On the other hand, people who travel widely learn to enjoy foods that are new to them, and familiar served in unfamiliar ways.

2.8 *Change in Food Supply*

Availability of food obviously exerts an influence on food consumption. Varied and more nearly adequate diets are found in countries where food production is high. In many newly developing countries the yield of crops per worker is low and diets are limited.

A plan in the United States known as the Food Stamp Programme was instituted in mid 1961 to make agricultural products available to needy families. The effect of the programme on the adequacy of the diet was assessed in selected areas. Before the Stamp Programme was in effect, one-half the families who later participated in the Programme had diets that met the Recommended Allowances. After they were in the Stamp Programme, two-thirds of the families had diets that met the recommendations. Eligible families who did not participate showed a negligible increase in number of diets meeting the Recommended Allowances.

2.9 *Trends in Food Habits in The United States*

The food consumption in this country, from 1910 when the first official records were kept to the present time, show certain general trends in eating habits. The change has been toward an improved diet. The commodities that have shown a general upward trend in per capita consumption are fruits and vegetables, and until something over a decade ago, dairy products. The lessened consumption of some items, including cream, butter, evaporated, and to a lesser extent whole fluid milk, turned the trend in the use of dairy products downward. It is the increased consumption of citrus fruits, tomatoes, and green and yellow vegetables that accounts for the upward trend in the consumption of fruits and vegetables. The foods that have taken a downward trend are potatoes and cereal products. However, the 1963 records show a shift in the consumption of potatoes with an increase over the 1957-1959 period. The consumption of meat and eggs has been variable. Since the depression of the early thirties, the per capita use of meat, fish, and poultry has moved upward. The trend in the use of eggs has been downward over the past ten years.

2.10 Food Supply for People of the World

Through the Food and Agriculture Organization of the United Nations information is gathered on the supply of food for the people of the world. There is not enough food at the time this is being written, nor was there enough at the first meeting of the FAO in 1945. In the intervening years, the quantity of food produced has increased but the world population has also increased. However, developments are promising. In 1945 food production per capita was not increasing as fast as the population; by 1949 the relationship was reversed with food production per capita in the lead. Dr. B.R. Sen, Director-General, Food and Agriculture Organization indicated in 1962, that taking the world as a whole, the rate of population growth in developing countries approached 2.5 per cent per annum, and the volume of agriculture was increasing at slightly less than 3 per cent per annum. Improved production within the countries and better global distribution of food has contributed to the more favourable supply. Much credit is due to the work of the FAO in helping the member countries to improve their farming practices.

The regions of population concentration and regions of large food production do not coincide. Total food production has increased in the past decade from as much as 20 per cent in some developing countries to as much as 50 per cent in others. Improvements in diets that would have been expected by the increase in food production have been, as Dr. Sen states, "largely nullified by population increase." Figure 2.1 shows the relationships between world population and agricultural production. The method used for identifying food deficit countries was computation of the national average food supply per capita and comparison of the amount of food energy, animal protein, total protein, and fat provided with minimum reference standards.

The projected trend of increases in population and food supply per capita indicate that in economically advanced countries agricultural output can be expanded easily, in less-developed countries agricultural production is increased with difficulty. The total increase in food supply between 1958 and 1980 sufficient to provide a minimum adequate nutritional standard, it is estimated, will require an annual average rate of increase of 3.5 per cent. Over the last several years the annual rate of increase in food production has been 2.9 per cent.

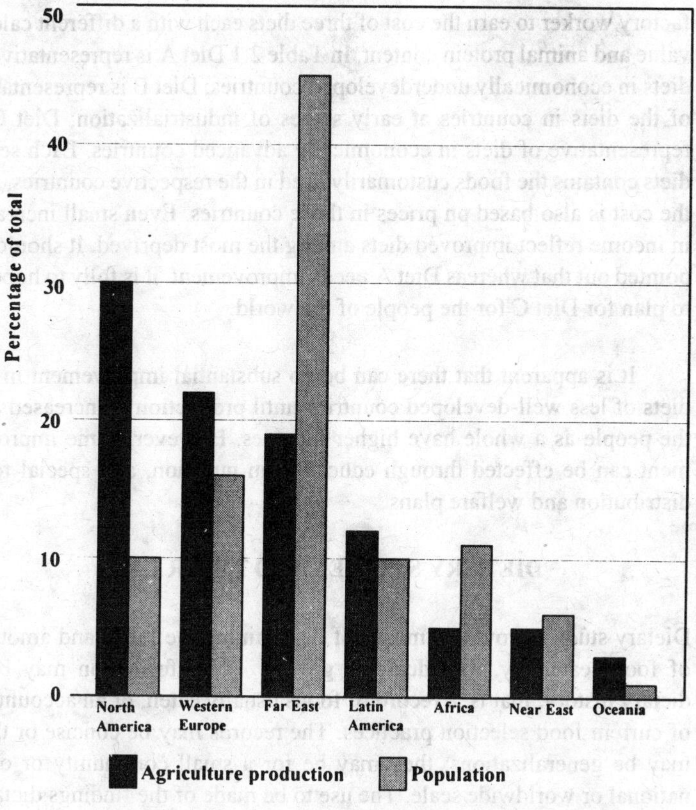

Figure 2.1 : Regional shares of world population and agricultural production, average 1957 to 1959. Production data based on price-weighted aggregates.

Despite the remarkable progress that has been made, problems remain — of how to increase productivity, how to distribute the world's food supply to avoid surpluses in one area and meager supplies in another, how to know what foods are needed and how much. There is a further problem with which the FAO workers must cope, and that is breaking through years of established traditions to persuade adoption of new and better methods of production and better use of the natural resources.

The cost of food in relation to income is a barrier to attaining a diet that meets a satisfactory nutritional level. In order to compare the cost of food relative to income in different countries, the FAO has estimated the income on the basis of the number of hours required for the average factory worker to earn the cost of three diets each with a different caloric value and animal protein content. In Table 2.1 Diet A is representative of diets in economically underdeveloped countries; Diet B is representative of the diets in countries at early stages of industrialization; Diet C is representative of diets in economically advanced countries. Each set of diets contains the foods customarily used in the respective countries, and the cost is also based on prices in those countries. Even small increases in income reflect improved diets among the most deprived. It should be pointed out that whereas Diet A needs improvement, it is folly to hope or to plan for Diet C for the people of the world.

It is apparent that there can be no substantial improvement in the diets of less well-developed countries until production is increased and the people as a whole have higher incomes. However, some improvement can be effected through education in nutrition, and special food distribution and welfare plans.

DIETARY STUDIES AND THEIR USE

Dietary studies provide a means of determining the kinds and amounts of foods eaten by individuals or groups. The information may be a dietary history, that is, a record of foods usually eaten, or an accounting of current food selection practices. The records may be concise or they may be generalizations; they may be for a small community or on a national or worldwide scale. The use to be made of the findings dictates to considerable extent the methods to be used in collecting the information.

Table 2.1 : Number of Hours 'Work Required for Average Factory Workers to Earn the Cost of Three Representative Meals for One Person for One day

	Diet A	*Diet B*	*Diet C*
Caloric value of diet	2100	2500	3000
Grams of animal protein	10-12	15-20	40-45
Country		*Hours of Work*	
India	2.1	2.9	4.5
Japan	1.8	2.3	3.8
Italy	1.1	1.5	3.2
Mexico	0.6	0.8	1.4
Sweden	0.4	0.6	0.9
Denmark	0.3	0.4	0.7
Canada	0.2	0.3	0.5
United States	0.2	0.2	0.4

Food consumption data serve widely divergent interests. For effective nutrition education programme, a knowledge of the food habits of the persons for whom the plans are being made is an essential prerequisite. Nothing can be more irritating than to be advised to follow a practice that is already in effect. In conjunction with the assessment of nutritional status, a dietary history of the usual food practices, together with current variations, permits a more advised interpretation of the findings.

3

Food Fads and Fallacies

3.1 Introduction

Fallacious beliefs concerning foods and their usefulness to the body are not new. However, with increasing scientific knowledge and improved means of communication, it might be expected that faddism would be less prevalent. The opposite seems to be true. The purveyors of half-truths or untruths "thrive by misinterpreting scientific facts" (3).

Concern for the influence of faddism on health has prompted the American Medical Association and the Food and Drug Administration to sponsor jointly national conferences on Medical Quackery. Food and Drug Commissioner George P. Larrick stated at the 1961 conference that, "the most widespread and expensive type of quackery in the United States today is the promotion of vitamin products, special dietary foods, and food supplements." An estimate made by the Food and Drug Administration is that 10,000,000 Americans spend $500,000,000 a year or $50 per buyer on nutritional nostrums.

Why does nutrition quackery thrive? It seems a paradox; the food supply is abundant and most people enjoy eating. Also, the purchase of food would be less expensive than the purchase of the special products and supplements. However, the faddist makes reckless promises of beauty, youth, health, and relief from pain and disease. It would seem that the interest of the people of the United States in science, and in health and medicine, should contribute to the vulnerability of the faddist. However, if there is considerable concern about health, it may be a favourable climate for quackery to be created. Milstead believes over-concern to be "one of the most significant factors in the increase in medical and nutritional quackery in recent years." Also the belief that science can accomplish magic may tend to make people gullible. The public is interested to learn of advances in science; the food faddist is ready and eager to satisfy that interest, but with misinformation.

Financial gain is the chief interest of the faddist. Extravagant claims are made for a product that has been concocted or for a dietary plan or health scheme sponsored in order to realize sales. The purchase of such nostrums and notions is an economic waste. Even more serious, however, is the delay incurred in seeking medical attention when it is needed.

Aside from the propagandists who have something to sell, there are many persons who have misconceptions about food. The misconceptions might be classed as folklore or "old wives" tales. The beliefs are handed down from one generation to the next; they are traditional, unquestioned. In this category, the admonition "to feed a cold and starve a fever" might be placed.

Some unsound beliefs are due to superstitions and fears. When traced back to their origins, circumstances were usually that illness had followed the eating of a certain food or food combination. Even though the illness was coincidental, the incrimination of the food as the cause persisted. An example of such an ill-grounded belief is the fear of combining seafood and milk in the same meal. There is no basis, of course, for the fear.

On the other hand, it may be wise to stop and consider practices that vary from those to which we are accustomed, particularly the habits of persons living in isolated areas. The beliefs they hold may well have a grain of truth. Centuries ago burnt sponge was recommended for simple goiter; later it was found to contain iodine, the element which if insufficient in the diet may cause a goitrous condition. The English fed limes to their sailors suffering with scurvy; later it was found that an insufficiency of ascorbic acid was the cause of scurvy and that limes contain that vitamin.

3.2 *Common Dietary Misconceptions*

A survey made by a committee of the American Dietetic Association indicated the following to be among the more commonly observed dietary misconceptions.

T*hat fruits, especially citrus and tamoto, are too acid to be handled by the body.* The acids in fruits are organic acids. With few exceptions after absorption they are oxidized to carbon dioxide and water, both of which compounds are readily eliminated. The inorganic residue remaining, after oxidation of the organic portion, is basic in reaction. Fruits are base-forming with the exception of prunes, plums, and cranberries, which contain organic acids (benzoic in prunes and plums and quinic in cranberries) that the body cannot oxidize.

That Garlic Cures High Blood Pressure. There is no evidence to indicate that garlic possesses either remedial or curative properties for elevated blood pressure.

That Beets Build Blood. Beets are not a good source of the nutrients recognized to be essential in blood formation. Protein and iron are the chief blood-building constituents. Copper also is essential for the formation of hemoglobin. Vitamins that function in the maintenance of a normal blood picture include folacin and vitamin B_{12}.

That Foods cooked in Alminum utensils will cause cancer. There is no basis of truth for such a belief. Foods cooked in aluminum utensils are increased in alumimum content, but little, if any, of the metal is absorbed from the gastrointestinal tract by man or animal. The small amount that may be absorbed exerts no ill effect.

That the following combinations of foods are poisonous : milk and orange juice or other citrus fruit; milk and fish. The belief is entirely erroneous. Such food combinations are eaten daily by many with no ill effects.

That raw cucumbers without salt are poisonous. There is no evidence to support this belief.

That a Good way to diet is to skip breakfast. Omission of breakfast results in lowered physical efficiency. Omitting some food at each of the regular mealtimes is a better plan than skipping some meals.

That Honey is not fattening. No particular food is fattening. Calories consumed in excess of need are deposited as fat. A tablespoon of strained honey (21 gm) is higher in caloric value than a tablespoon of sugar (12 gm), 65 and 45 Cal., respectively.

That meats gives you strength. Meat has a high satiety value. The extractives give a flavour that is enjoyed and the protein is effective in delaying the onset of hunger sensations.

That fruit juices do not contribute calories to the diet. Fruit juices contribute Calories to the diet as follows per cup: lime, 65; lemon, 60; orange, 110; pineapple, 135; and prune, 200.

That toast has fewer calories than bread. The caloric value of bread is unchanged by toasting. Water is lost in the process and part of the starch is dextrinized, neither of which changes the energy value of the bread.

That vegetable fats and oils can be used in nay quantity and are not fattening. Vegetable fats and oils are high in caloric value, and are comparable to animal fats.

That adults need no milk. Without milk or a milk product in the diet, it is difficult to supply the recommended calcium allowance of adults. In addition, dairy products provide in significant amounts protein of high quality, phosphorus, vitamin A, and riboflavin.

That skim milk has little nutritive value. Skim milk is whole milk with the butterfat removed. It remains a superior food, even though low in fat and vitamin A content.

That pork liver is less nutritious than beef or calves liver.. The composition of the three meats is similar in energy value and protein content. Pork liver in higher is iron and thiamine content and lower in vitamin A as shown below:

100 Gram Edible Portion Cooked

	Iron,	*Vitamin A,*	*Thiamine,*	*Riboflavin,*
	mg	IU	mg.	mg.
Beef liver	8.8	53,400	0.26	4.19
Calves liver	14.2	32,700	0.24	4.17
Pork liver	29.1	14,900	0.34	4.36

That yogurt is an aid in relating youth and beauty. There is no evidence to indicate that yogurt or other fermented milks are wonder foods. Yogurt is whole milk that has been inoculated with the microorganism *Lactobacillus bulgaricus*. It is consumed by some because the flavour is enjoyed. In certain cases it has been recommended to patients as a means of lessening the incidence of the normally occurring putrefactive bacteria in the intestine.

That natural foods are the only ones that are safe for the consumer. The safety of additives to processed foods is guarded by the Food and Drug Administration; nutrients are added to some processed foods to enhance the nutritive value.

That white-shelled eggs are more nutritious than brown. The nutritive value of an egg is not related to the color of the shell; that is determined by the breed of the hen.

That large amounts of gelatin dissolved in water and taken as a food supplement will strengthen fingernails. Fingernail formation apparently is influenced by a number of factors including state of nutrition, endocrine state, disease, and environment. Gelatin is a pure protein. Whether protein of itself, in any form, will improve fingernail condition is questionable.

That water is fattening. Water has no caloric value and therefore cannot be converted to body fat.

3.3 Detecting Faddism

There are certain characteristics that help the unsuspecting public to discern truth from fiction. Mitchell (9) has described the characteristics of misinformation and the perpetrators of it as follows:

> Earmarks of quackery and misleading propaganda should be familiar to the layman. Extravagant claims for cure or relief of various complaints are always to be doubted. Vague labeling which offers no figures or exact information to live up to is one way by which these promoters have avoided the teeth of the law. Labels are never so extravagant as detached advertisements or literature because the laws are stricter about labels.
>
> Promoters of most of the fad diet systems or health institutes have personality, poise and persuasion. They are convincing speakers with a smattering of scientific knowledge. This lends an air of plausibility which masks overstatement, understatement, and sheer misinterpretation. They mix the true and the false or misleading to suit their needs, to sell their products, or to scare their victims into spending money for a course of lectures to cure than of imaginary ills. They use scientific words and phrases to cover false ideas. They accomplish most by personal appearances but they also work through the mails, the radio, the newspapers, handbills, newsstand publications, and books with arresting titles.

And it should be added, they operate door to door.

3.4 Myths about Food and Nutrition

There are four myths that the Food and Drug Administration finds to be almost universally propounded by the person with a product or a nostrum to sell.

The number first myth is that *all diseases are due to faulty diet*. The premise is that it is almost impossible to consume an adequate diet from foods, hence the promotion of the product to be sold. "The product in question usually contains a long list of ingredients, often including some which are supposed to posses secret benefits which nutrition scientists have not yet discovered." Janssen (6) pointed out that "the promoter with the pill that '*has* everything' will generally also claim that it can 'do everything' — in other words, that it is a cure-all and the answer to every health problem."

The second myth is that *soil depletion causes malnutrition*. There is no scientific evidence to support this statement. An impoverished soil diminishes the quantity of the crop produced rather than the quality. Some slight variations in the nutritive value of crops have been related to soil fertility, however, of more importance is the genetic strain of the crop.

The only recorded instance of a relationship between human nutritional status and soil composition is that of the incidence of endemi goiter in regions of low iodine content of the soil. The deficiency c iodine is taken care of by the use of iodized salt in the United States and some other countries also.

The third myth commonly promoted is that of *overprocessing*. The fact is, that some methods cause some loss of minerals and vitamins but most of the nutritive value is retained. This misinformation is employed to support the sale of certain cooking utensils or food supplements. Furthermore, food consumption data indicate that the supply of nutrients per capita is at an all time high (5).

The fourth myth is concerned with so-called *subclinical deficiencies*. Janssen explains the implications of this myth as follows: "According to this theory, anyone who has a tired feeling or an ache or pain in any part of the body is probably suffering from a subclinical deficiency and needs to supplement his diet with various concoctions. The food-fad promoters teach that practically everybody is in this category because of our depleted, over processed, and poisoned food supply." Such cymptoms are universal and may result from many causes. Symptoms of dietary deficiencies should be diagnosed by a competent physician.

A false statement that influences old persons in particular is that, as a person grows older "he develops an increased or unusual requirement for some nutrient or food that is not supplied by the ordinary diet." Actually the energy requirements *decrease* with age. In fact, the Recommended Allowances for none of the nutrients increases with age. To the contrary, for women in the 55 to 75-year group, the recommendation for iron is less than for the younger adult group; for riboflavin and niacin, the recommendations are less for both men and women, and less for thiamine for men.

3.5 Means of Deception

Some of the kinds of products and practices that are avenues for deception include:

Mail Order Products Responsible mail order companies give full value for purchases made. However, the Food and Drug Administration warns that exaggerated claims are prevalent in mail order literature on health products. The Postmaster General has warned that quackery in this field is at an all-time high. This warning applies particularly to mail promotion of nutritional products.

"Health Lecturers" The lecturer usually attacks well-known foods or products and has for sale a special line of health foods or special utensils.

Books and Literature There are many good books on nutrition and health on the market. However, "any book that advocates a quick 'cure' for a serious illness such as cancer or arthritis is certainly questionable". Also there are books on the market that are inaccurate in content and filled with misinterpretations written in an authoritative style.

Doorbell "Doctors" We are warned to beware of house-to-house salesmen of vitamin and food supplements. They may "attempt to discuss your health and dietary problems and to give advise regarding them. They are salesmen, not doctors".

3.6 Means to Counter Misinformation and Fraud

Stronger laws, stronger law enforcement, and public information and education were suggested by Commissioner Larrick of the Food and Drug Administration as three means of alleviating the influence of misinformation and fraud.

Public service groups that can lend assistance in detecting deception are: Better Business Bureau; Department of Investigation, American Medical Association, 535, No. Dearborn St., Chicago 10, Ill.; U.S. Post Office Department, Office of General Counsel, Washington 25, D.C. for queries about mail order products; Food and Drug Administration, U.S. Dept. Health, Education, and Welfare, Washington 25, D.C.

Additional sources of authoritative information about food and nutrition for the professional and lay person are: agricultural and home economics extension services located at state universities and with country offices; food and nutrition departments in home economics in colleges and universities; state departments of health; Agricultural Research Service, Human Nutrition and Consumer Use Division, U.S. Department Agriculture, Washington 25, D.C.; American Dietetic Association, 620 North Michigan Ave., Chicago 11, Illinois; American Home Economics Association, 1600 Twentieth St. N.W., Washington 9, D.C.; Council on Foods and Nutrition, The American Medical Association, 535 North Dearborn St., Chicago 10, Ill.; Food and Nutrition Board, National Academy of Science — National Research Council, 2101 Constitution Ave., Washington 25, D.C.

3.7 Protection of the Consumer

We are fortunate in this country to enjoy certain assurances by law that our food supply is wholesome. Also we may feel confident when making a purchase that the product conforms to the label description. Some of our protection is through federal laws, some by state laws, and some through local regulations. The federal regulations cover products that pass through interstate commerce, the state laws are effective within the state, and local regulations such as country and city apply to those circumscribed areas.

Important among the federal regulations as prescribed by the Food, Drug and Cosmetic Act of 1938 is that of accurate labeling of products. Foods bearing labels that are false or misleading are prohibited to be shipped from one state to another or imported from another country. The federal law in addition imposes certain specific requirements for the labels of foods intended for special dietary uses in order to inform the purchaser of the value of the food for such uses. It is required that the vitamin and mineral content and other dietary properties be declared on the label. Figure 3.1 is a copy of a label from a cereal preparation forbabies. The federal law also makes a special requirement that foods intended for dietary management of disease or the control of body weight carry labels stating the percentage by weight of the protein, fat, available carbohydrate content and the caloric value of a specified quantity of the food. Figure 3.2 is the label from a can of grapefruit prepared for use by persons on special diets limited in sugar content and persons on low-caloric diets.

	Approximate Analysis %
Carbohydrate	44.9
Protein (N×6.25)	35.0
Moisture	7.0
Minerals (ash)	7.3
Calcium	0.90
Phosphorus	0.90
Iron	0..30
Copper	0.001
Fat	4.0
Crude fiber	1.8
Calories per oz	105
Tablespoonfuls per oz	12

Each ounce contains not less than 0.3 mg thiamine, and 0.1 mg riboflavin and supplies the above percentages of the minimum daily requirements of nutrients.

Fig. 3.1 : Label from a special cereal preparation for babies. (Courtesy of Pablum Products, Division of Mead Johnson and Company, Evansville, Indiana.)

The regulations of the state laws vary from state to state. Most of them cover mislabeling of products and provide for certain sanitary regulations of products manufactured and put on the market within the state. Local laws apply for the most part to control of the quality and sanitation of milk and milk products on the market and to the sanitary condition of local eating establishments.

No Sugar Added

Typical Analysis of Food Values

Available Carbohydrates 8.2%

(total Carbohydrates minus crude fiber)

Protein 0.7%

Fat 0.1%

Minerals 0.5%

Sodium per 100 grams, 3 mg.; per ½ cup, 3 mg.

Available Calories: per 100 grams, 37; per ½ cup, 42

For Dietary Purposes

Prepared from fully matured, carefully selected, sun-ripened Florida Grapefruit; Grapefruit Juice; and non-nutritive Sucaryl (Cyclamate Calcium, Abbott), 0.1% Cyclamate Calcium is a non-nutritive artificial sweetener which should be used by those persons who must their intake of ordinary sweets.

Fig. 3.2 : Label from a can of grapefruit for persons on diets limited in sugar content (Courtesy of Tillie Lewis Foods, Inc. Stockton, California.)

The Federal Trade Commission is another agency that assists in protecting the consumer through the control of false and misleading advertising. The Commission can take action against claims that show a tendency to mislead or could be construed as deluding the public through advertising by way of radio, television, public lectures, pamphlets, books, newspapers, form letters, and other types of communication. The Commission can act, however, only if the false or misleading matter can be classified as commodity advertising.

The protection afforded the consumer by governmental agencies does not relieve the individuals of their responsibility in attempting to discern between fact and fiction.

4

The Flavour and Aroma of Food and Food Additives

4.1 Introduction

Flavour is the subtle and complex sensation that is the source of much of the delight man finds in food. To both connoisseur and ordinary man, flavour is of utmost importance in regulating preferences. It is the difference between a cheap wine and the most expensive, the highest grade butter and the lowest, rancid cookies and fresh.

A *food additive* is any substance not naturally present in a food but added during its preparation and remaining in the finished product. It also includes those substances which are naturally present but their concentration is increased by fortification. However, since salt, sugar and vinegar have been used for centuries they are not considered as additives. Today many compounds are added to the foods to protect against spoilage, enhance flavour, improve nutritive value and to provide some special property to food.

Due to toxic nature of some food additives it was thought desirable to ensure the safety of new additives. The toxicity of pesticides for insects alarmed some consumers who believed that surely these compounds that kill insects might be equally lethol for themselves. It was thought fit to provide some legal safeguards.

4.2 Legal Safeguards for Food-additives

Agene, nitrogen trichloride, was used as a bleach for flour for many years. Bleaching improves the baking quality of flour in a short time and produces a flour comparable to the more expensive aged product. Agene was used in very low concentration but it was shown in the late 40's that flour bleached with this agent or bread baked from this flour caused running fits in dogs and convulsions in rabbits, ferrets, mink, and cats—a report which led to an almost hysterical distrust of all food additives in a small segment of our population. The animals adversely affected were fed diets composed principally of the flour or bread. The toxic compound is produced from the action of nitrogen trichloride on methionine to form methionine sulfoxamine. It has never been demonstrated that harmful effects are caused by this flour or bread in man although a number of experiments have been conducted to test the possible toxicity. So, agene was removed from the list of permissible bleaching agents.

After much debate and a number of hearings and bills, the Miller Amendment was passed in 1954 and the Food Additives Amendment in 1958. The Miller Amendment covers spray residues and the use of pesticides; it attempts to establish procedures that will prevent any possible harm to consumers. The Food Additives sets up requirements for the demonstration of the safety of a substance before it can be used in foods.

The Food Additives Amendment uses the following definition:— "'Food additives' include all substances not exempted by section 201 of the act, the intended use of which results or may reasonably be expected to result, directly or indirectly, either in their becoming a component of food or otherwise affecting the characteristics of food." The container is not considered a food additive but if some of the substances present in the container migrate into the food, then they become food additives and subject to the law.

A list of additives that are already in use and appear to have been adequately shown to be harmless has been prepared. No further testing is required with these compounds that are considered permissible. The list includes preservatives, buffers and neutralizers, nutrients, non-nutrient sweeteners, colouring agents, stabilizers, emulsifiers, and others. Although salt, sugar, and vinegar are not included, they are permitted. A list of permissible flavour substances is available. Other additives cannot be used unless permission is granted by the Food and Drug Administration. A petition for use must include a thorough review of the substance, its stability under all conditions of possible use, and a complete report on animal tests that demonstrate its safety. Methods for the detection of the additive or its degradation products if they are formed in the food must also be given. The Food and Drug Administration guarantees protection of trade secrets. Only those facts about the substance that are required for its control will be revealed.

4.3 Why Use Food Additives?

At a meeting of the Institute of Food Technologists in May, 1958, Harold Schultz is reported to have said of food additives that "some are used to colour foods, others to bleach them: some add flavour to food, others remove flavours; some make foods firmer, others soften them; some keep foods dry, others keep them moist; some thicken foods, others keep them from thickening; some produce foams, others prevent them; some make foods acid, others make them alkaline, still others suspend particles; some are added to increase the mineral ions, others to remove them; some are oxidizing agents, others are reducing agents; some hasten chemical changes, others retard them." This is a formidable list of functions although sometimes one food additive has more than one. It is a list that illustrates the needs of the food industry in its efforts to produce foods with sufficient shelf life to allow distribution over a wide area and with sufficient appeal to the eye and taste of the consumer to arouse and satisfy his appetite. These food products must nourish the eater well; but before they can nourish him, they must first attract his purchase. Justifiable uses of food additives have been described in some detail by the Food Protection Committee in Publication 398. "The Use of Chemical Additives in Food Processing." They are:

(1) "maintenance of nutritional quality, such as by the use of antioxidants;"

(2) "enhancement of keeping quality or stability, with resulting reduction in food wastage, through the use of antioxidants, antimicrobial agents, inert gases, meat cures, etc;"

(3) "enhancement of attractiveness of foods by means of colouring and flavouring agents, emulsifiers, stabilizers, thickeners, clarifiers, and bleaching agents. (That appeal to the eye as well as the palate is not a mere whim of the food producer, is recognized by all nutritionists, since proteins, vitamins, and minerals serve no purpose unless eaten.);"

(4) "providing essential aids in food processing. Agents which function in this capacity include acids alkalis, buffers, sequestrants, and various other types of chemicals."

4.4 *Methods of Demonstrating Safety*

The current methods of demonstrating the safety of a chemical compound or crude extract to be used in food production are fairly well standardized. Those compounds which are not normally present in some foods are fed in graded amounts to rats for two years and to dogs for one year. The control animals are fed complete diets devoid of the additive. The level of feeding at which injury can be demonstrated is determined.

The success or failure of this method depends on recognition of "injury." A considerable amount of attention has been devoted to this effect. The Food and Drug Administration uses the distinction between toxicity and hazard set forth by the National Research Council's Food and Nutrition Board: "Toxicity is the capacity of the substance to produce injury; hazard is the probability that injury will result from the use of the substance in the quality and in the manner proposed." Since evaluation of injury depends on subjective judgment when changes are slight, the FDA requires a complete description of methods used. "No effect" must be supplemented by a description of the conditions under which there is "no effect." The decision of the safety, toxicity, or hazard of a chemical additive rests with this organization.

For some substances it is possible that use in small amounts proves safe but in very large amounts hazard exists. Even a substance such as a sodium chloride, commonly considered completely safe to add to foods, can cause injury when the amount far exceeds that encountered in foods. With these compounds a limit is established by adding a margin of safety to their toxic level. This margin must include consideration of the difference in effect between man and test animals, individual variation, cumulative effect, and the possibility of the compound occurring in other items in the diet. Here again judgments rather than measurements are used and much debate is possible.

It is unlikely that most chemical additives will be chemically pure compounds. FDA requires that methods for the detection and estimation of impurities be developed and presented with the petition for permission to use. Although a pure substance might be safe, the impurities carried along might be injurious.

There is also a possibility that a compound might be safe as such but during the processing of a food, it could cause a change in some component to form a poisonous compound. The toxic factor in flour bleached with agene proved to be a derivative of methionine rather than the agene itself. It is therefore required that a compound not only be demonstrated safe, but also safe under the conditions of its use. Permissible chemical additives are restricted to those particular foods or processes which meet these requirements.

4.5 The Sensation of Flavour

Flavour is a combination of taste, smell and feel. In the mouth and pharynx are many taste buds capable of detecting sweet, sour, salty, and bitter. In the nose are olfactory endings that can detect a huge number of different odours. The "mouth feel" of a food is likewise part of its flavour — whether it is smooth or rough, tender or tough, unctuous so that it clings to the tongue and roof of the mouth or watery so that it slides down readily. "Aftertaste" is a quality for the most part that is the combination of all of these sensations after the particle of food has been swallowed. However, even in aftertaste the sensation of feel has specific importance since the stickiness or greasiness of the small amount remaining in the mouth and on the teeth contributes to the general aftertaste.

Taste

This sense is detected through the solution of soluble compounds in the saliva or in the food juices and the contact of those dissolved compounds with the taste buds. The commonly accepted theory (although not without challengers) of taste is that there are four primary tastes that can be detected: sour, sweet, salt and bitter. The areas in which these tastes are detected overlap, but the sensation of sour is most readily detected on the sides of the tongue, salt on the sides and tip, sweet on the tip, bitter at the back of the tongue and on the pharynx. The taste buds are present in greatest number in the vallate papillae, the tiny nipple-shaped elevations distributed in a V on the tongue. They are also present in papillae on the rest of the tongue, the soft palate, pharynx and epiglottis. The taste buds are composed of a number of cells arranged in a tiny well around a nerve ending.

A compound to be tasted must occur in solution or dissolve in the saliva. The solution seeps into the taste bud and the compound stimulates the nerve ending. An impulse is transmitted along a nerve to the brain and we recognize a taste. Then more saliva washes the solution out of the tiny well.

Taste thresholds are measures of the sensitivity to a given taste in terms of the detectable concentration. A common method for measuring the threshold of, say, salt, is to give a number of samples with varying concentrations of sodium chloride, a number of samples of pure water, and sometimes of other compounds with other tastes. The subject attempts to detect the taste present, rinsing the mouth after each attempt. The lowest concentration consistent for an individual is his taste threshold for a particular substance. The low levels at which tastes can be detected is truly amazing. In a group of students almost everyone could detect 0.087 per cent sodium chloride and 0.4 per cent sucrose. Considerable variation between individuals occurs — a low threshold for one taste not always accompanied by a low threshold for the others.

Influence of Chemical Constitution: Taste depends on a number of factors, the most important of which is chemical constitution. Saltiness is a property of electrolytes, and the halides particularly. In order of their saltiness, the following ions are detectable: Cl^-, Br^-, I^-, SO^-_4, NO_3^-. The cations also influence the taste. Na^+ and Li^+ salts tend to have salty tastes while K^+, salts have considerable bitterness.

Sweetness is found in a number of organic compounds. Alcoholic hydroxyls tend to endow a compound with sweetness, but other groups are sometimes effective. Saccharin, for example is a totally different compound from sucrose, possessing no hydroxyl groups, but rather a sulfonamide. Sucaryl with 30 times the sweetness of sucrose is also a sulfonamide.

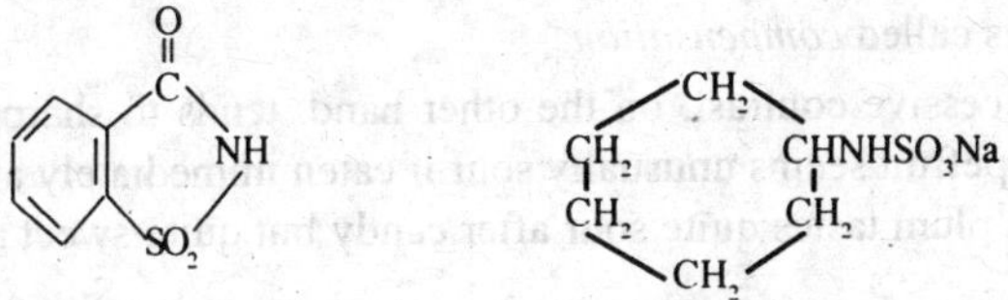

Saccharin   Sucaryl Sodium

Generalizations concerning groups that make for sweetness in organic compounds are difficult.

Bitterness is a property of some organic and inorganic compounds. Some of the alkaloids such as quinine and brucine are exceedingly bitter, and NH_4^+, Mg^{++}, and Ca^{++} are also bitter.

Sourness is a property of the hydrogen ion; its concentration is of primary importance in determining whether or not the sensation of sour is detected. The acids that occur in foods are organic with relatively low ionizations and consequently relatively low hydrogen ion concentrations. Added to this is the common occurrence of a salt of the organic acid in the food, which further lowers the ionization by common ion effect. Some workers believe that aside from the effect on ionization, the anion likewise has an influence on the perception of sourness. Research upon rats concerning their nerve response to acids shows variation in response when hydrogen ion concentration is constant.

Influence of Other Factors: Taste is also influenced by temperature, texture, and the presence of other compounds. Mackey and Valassi measured the taste thresholds for the four primary tastes in water as well as in tomato juice and in egg-milk custard prepared as liquid, gel and foam. They found that the primary tastes were harder to detect in gels than in liquids and that the foams were intermediate.

The effect of one primary taste on another is well–known to everyone who cooks. Salt tends to decrease the sweetness of sucrose, so that candy to which salt is added has a richer, less sweet taste than that without salt. Sugar likewise can tone down and round out saltiness. Many a cook has salvaged the mashed potatoes that were too salty, by adding a small amount of sugar. The effect of sweet or sour such as sugar on lemon juice on vinegar, is also important. Any food quite sour and yet quite sweet has a rich, full flavour that is lacking when either sourness or sweetness is present along. Salt also tones down sourness of food acids. This toning down of sensation when two tastes are presented simultaneously is called *compensation*.

Successive contrast, on the other hand, tends to sharpen a sensation. Grapefruit seems unusually sour if eaten immediately after a sweet cereal. A plum tastes quite sour after candy but quite sweet after grapefruit.

An interesting phenomenon has been discovered in relation to the taste of a few compounds. Sodium benzoate tastes quite different to different individuals and the threshold at which it can be detected is very different. Phenyl thiocarbamide (PTC) has been extensively studied: "taste blindness" to this compound is inherited according to Mendelian ratio. If both parents are nontasters, the children do not taste PTC. Women tend to have a slightly lower threshold than men do for this compound. Taste blindness in women occurs in 22.2 per cent and in men 25.9 per cent.

Fatigue for taste does not occur rapidly and some physiologists do not believe it ever occurs. However, it is a common experience to notice a diminution in the taste of a food as one continues to eat it. Some experiments indicate that fatigue occurs for one taste but the others are not affected or are even enhanced.

Odour

We can detect differences in the odour of thousands of compounds impinging on the olfactory nerve endings. The sensation is experience, of course, only when the nerve impulse is transmitted to the brain and recorded there. Our remarkable ability to remember odours and the ability of the brain to receive and detect differences in odours cannot be adequately explained at present. Many people can recognize odours that they have experienced only once, many years before. The author once heard a 70 years old woman proclaim that a noxious odour must be a stinkhorn mushroom since she remembered smelling one as a girl. The hoots of derision were quieted when a stinkhorn was found to be the source of the odour, Many individuals have experienced this remarkable recall for odours.

The aroma of food is recognized as a tantalizing and delectable prelude to eating but not always as an invaluable part of flavour. We have only to suffer loss of our ability to detect odour through congestion of the nose, to realize what a contribution it makes. Food tastes salty, sweet, sour, and bitter when we have a bad cold, but how flat and uninspired it is without odour. All the subtle nuances of the flavour complex are gone.

The olfactory mocous membrane is located in the upper part of the nasal cavity on portions of the turbinates and septum, and contains olfactory nerves imbedded in a special epithelium. Glands are present that secrete fluid. In order for a chemical compound to possess an odour it must vaporize and pass into the nasal cavity. It is not known whether the molecules dissolve in the fluid on the lining and the solution comes in contact with the olfactory cells or whether the hairs of the cells penetrate the mucous and come in contact with gas. The nerve then transmits an impulse to brain. The detection of an odour during breathing occurs through the diffusion of the gas into the still air of the upper cavity. The olfactory endings are not located on the epithelium in the direct path of the inspired air. When we detect a flavour in a mouthful of food, the odorous compounds change to gases and diffuse through the pharynx into the nose. As we raise the food to our mouth, we inhale its odour.

The sense of smell varies considerably between individuals and is, of course, much less keen in man than in many other animals. Nevertheless, it is still remarkably keen for some compounds. Vanillin can be detected by most individuals at a concentration of one part in ten million.

Fatigue for odours occurs very quickly. We have all observed how strong an odour of a food such as coffee or meat will be when we first come into a building or room where it is being prepared and how quickly we loose that sensation. Students who do not work in chemistry, often complain of the odours of the laboratory while those of us who work there constantly, do not detect them unless the concentration suddenly increases or a new gas is formed. People have been overcome by hydrogen sulfide, which has an unusually strong odour, after fatigue of the olfactory sense was so depressed that they forgot the concentration was high.

Although fatigue for odours occurs rapidly, it is selective. We may not be able to smell the odour of a compound after a short period of time, but the sense of smell for another odour is unimpaired.

Testing Sense of Smell or Aroma: Elsberg devised a method for testing acuity of smell. A flask is filled with cotton and between the layers is a toothpick, cotton, wrapped, that has been dipped in a solution of some essential oil or odorant, often a pure chemical. Air is admitted to the flask in measured amounts and the vapours sniffed. The amount of air required to perceive the odour and also the concentration of the solution on the cotton give a measure of the acuity of the individual to that particular odour. Some individuals have high acuity of smell but cannot readily recognize odours. Thus, another part of the test often introduced is proper description of the odour perceived.

Some physiologists are now studying taste and odour responses in animals and attempting to obtain data that will make possible an understanding of these phenomena. Since the impulse that travels over a nerve is electrical, measurement of the potential of a nerve shows the effect of a stimulus. It is possible in a living animal to make measurements on the large nerve bundle that serves many taste buds, on an isolated fiber showing the response of several taste cells in one taste bud, or by inserting microelectrodes to make measurements on a single taste cell. Olfactory response is more difficult to measure although record of the activity of the olfactory nerve can be made. It is also possible to measure the change in potential in the presence of an odourant across olfactory tissue after it is excised from the animal. None of these studies is easy, but it is here that we can hope to find better understanding of the physiology of taste and odour.

Feeling

Some compounds not only stimulate the olfactory ending but also that of the trigeminal nerve—a nerve with endings in the skin of the face, tongue and teeth, giving a general sensation. Ammonia, a common substance that stimulates this nerve, has an odour for us but also another sensation described as "tingling" or "sharp."

Other "feel" factors contributing to the over-all sensation of flavour are the hot, burning effect of some peppers and spices as well as the coolness of peppermint. The texture of the food is also important: grainy fudge tastes sweeter but not as rich as creamy fudge. Crispness and stickiness are other sensations of mouth feel. Some wines contribute to flavor by seeming to fill the mouth and cover the pharynx immediately after a small sip.

Blends

A blend of sensations occurs with most foods so that sorting out all of them is difficult. This is often highly important to the satisfaction that a superbly delectable dish affords. Many foods have a slightly sour note that most people do not separate as "sourness" but simply as part of the delightful flavour. Broccoli, for example, has a volatile acid which is part of its fine flavour and most honey contains appreciable amounts of acid. Occasionally elements of flavour appropriate and desirable in some foods are highly repugnant when present in others. For example beer has a sour, horsey aroma and some ripened cheeses have an element of rancidity in the odour but, because it is expected and blended, it is pleasing. In the Arthur Little Flavour Laboratory scientists have tried blends of many ingredients, pure compounds and extracts, and conclude that in blending various reactions may result. With two odourants: (1) some of the major notes of each odourant may be suppressed; (2) all of the major notes of each odourant may be suppressed; (3) some of the major notes of the odourant and suppressed but none of those of the other odourant; (4) a complete blend takes place with the formation of a new odour; or (5) a partial blend occurs with a new odour formed but some characteristics of each odourant retained. There is still much to learn about blending, masking, and enhancement when two or more odourous compounds are mixed.

Appropriateness is of great importance in our enjoyment of flavours in foods. An onion flavour may be delectable in a stew or soup but objectionable in a custard. We become conditioned to expect certain sensations from certain foods and while a slight variation is titillating, a completely unexpected taste is unacceptable. Most of us have a "sweet tooth" and relish anything sweet. But while we may find that chicken prepared with the addition of a small amount of thyme is exceedingly interesting because the flavour departs a little from that which we usually experience, chicken prepared with a strongly sweet sauce may be rejected as improperly prepared and unappetizing.

Aftertaste is part of the sensation of flavour of some foods. It is particularily common with foods that leave a residue in the mouth after swallowing. Sticky foods such as syrup or greasy foods such as meat fat tend to leave a residue coating the mouth for a short time after swallowing and all or part of the sensations which are part of flavour, continue to be experienced. The persistence of the flavour is highly desirable in some foods, although usually we do not enjoy the sensation if it persists too long. Many people do not eat raw onions because of the persistent aftertaste. The volatile disulfides present are absorbed by the blood and are then secreted in the saliva and exhaled in the breath, so that the odour may persist for many hours.

4.6 Control of Flavour and Aroma in Processed Food

Control of flavour and aroma in processed food is of utmost importance in determining the quality and selling price of the finished product; this is likewise true of simple cookery operations. Many factors, all of which must be considered, influence the flavour. The quality of the ingredients has, of course, a considerable effect of the product. Off-flavour ingredients cannot produce an item with excellent flavour. Conditions of processing must be carefully controlled. Thus in roasting coffee or coca beans the temperature and length of time roasted must be controlled in order to insure flavour of the right quality. Avoidance of contamination by flavourful compounds during processing or storage must be watched. Thus the lacquer lining of cans cannot contain soluble flavoured substances. Many foods such as fats, and baked goods, pick up odours of other foods readily and must be adequately protected from contact with these odours. Foods must also be protected from contamination by bacteria and moulds and stored under conditions where these microorganisms cannot grow rapidly. Moulds produce compounds that impart a musty of sour flavour which is completely unacceptable.

In many food industries, tasters are employed to grade the product and often the raw materials. There are usually no chemical tests that can differentiate the subtle differences between a quality product, an ordinary one and one that is off-flavoured. With a little experience, some tasters develop remarkable ability to differentiate products. Some can readily detect flaws in the processing operation and spot the location of the trouble by taste. In many industries these tasters are of great value to the company since flavour is one of the bases on which the consumer buys the product.

Good flavour in a food product is desired by everyone who works in food production whether a housewife or the director of a nation-wide industry. Caul has analyzed the pattern of good flavour as the following sensations: "(1) an early impact of appropriate flavour; (2) rapid development of an impression of highly blended and usually full-bodied flavour; (3) pleasant mouth sensations; (4) absence of isolated unpleasant notes; and (5) anticipation of the next mouthful."

4.7 Measurement of Flavour

The sensations of flavour are such complex reactions that it is impossible by any simple chemical or physical test to measure them. An appreciation of flavour depends on a human being and his reactions to the actual act of tasting. Since the flavour of food products is so important in determining their commercial value, many methods for measuring flavour have developed. These have been reviewed at various times. A recent review of the methods and their implications is in a symposium in Food Technology for 1957.

Expert tasters are employed in some industries, particularly for wine, whiskey, tea and coffee and spices. The expert taster has usually grown up in one industry and is of value only to this industry. Through interest and opportunity he has developed a "discriminating palate" for the food of his company. There is no indication that an expert taster has a keener sense of taste than the normal individual, but rather that he has years of training.

Other methods use a panel of tasters; sometimes these are individuals with some training, sometimes they are workers in the industry with sufficient interest to be willing to serve on a panel, and occasionally where preference is desired, they are large numbers of consumers. The panel may indicate its preferences of judgment of quality by scoring a food on some well defined qualities. Often a numerical scale is used so that the scoring of the individuals can be added readily to give a composite score. Scoring has been widely used in meat, bakery, and milk products and in recent years with dehydrated foods.

Difference tests are sometimes used in an attempt to get a more precise and reproducible test of flavour in foods. A trained panel is required for these tests. Boggs and Hansen have summarized the various methods of difference testing. Either a pair of samples or a triangle in which two of the samples are identical and the third is to be separated are presented to the judges. Sometimes in a triangle a standard is submitted and one of a pair is matched to it. In other tests the judges only know that two samples are identical.

The dilution test is a type of difference testing. A sample is presented to the judges and then other samples that may or may not contain the unknown at a definite level of dilution are offered. The method was first applied to the detection of dried egg in fresh egg. The higher the quality of the dried egg the greater the difficulty of detecting it. In order to obtain a high degree of agreement, it would be necessary to have a high per cent of dried egg present. A poor quality product would be detectable at a low concentration or a high dilution.

Ranking is used with a flavour panel and with either a numerical or alphabetical evaluation of one property. A series of samples are supplied to each member of the panel and he arranges them in the order of increasing or decreasing quality of the characteristic. A high degree of agreement between members of a panel occurs when the samples differ to a marked degree; but when differences are subtle, agreement is not as common.

The *flavour profile* is a method for evaluating a flavour by describing it either as a whole or by characteristics. The vocabulary by which these characteristics are described are comparative terms such as "rubbery" and "eggy," or, when possible, they are referred to a compound—phenyl acetic acid for "horsey." Feeling effects are described on the basis of the mouth, throat or nose reaction, such as "throat burn," "puckering," and "cooling." The reactions are broken down into (1) character notes, (2) order of appearance, (3) aftertaste, and (4) amplitude. The character notes are the producing sensations, often exceedingly difficult to distinguish in a blend. The order of appearance may appear at first sight to be of little importance or even absent. But close attention to flavour during tasting will reveal that many flavour are not perceived as a whole but rather as a series of sensations. These come very rapidly but can be separated. Coffee bitterness is not perceived until after the other flavours have been experienced. Usually an unpleasant note should not be either the first or last sensation of the flavour. Aftertaste has been mentioned before, and its importance in the taste complex is readily recognized. "Amplitude" is the term used to express the total effect of flavour and aroma. The fitness of the flavour, the broad aspects of all characteristics is included here. It requires some understanding and experience to judge amplitude. It is rated as very low, medium or high. Examples of amplitude can show the meaning of this term. Tomatoes picked green have low amplitude, those grown in a hot house are medium, and the vine-ripened are high. Potatoes, boiled but unseasoned, have low amplitude; salt added at the table improves the amplitude by blending the flavour but salt added in preparation gives high amplitude.

Any evaluation by a flavour panel is subject to error. Many of the precautions necessary to the satisfactory use of a panel have been explored and described. It is essential that the members work alone in quiet, restful surroundings. Usually air-conditioned booths are used so that even the temperature and the humidity can be controlled. To avoid influencing one another, the panel members do not discuss their reactions with each other. Even the day of the week and the hour of the day has an effect. Mitchell found that the sensitivity of a duo-trio test where two samples out of three are matched, is greatest on Tuesday with Friday also better than Monday, Wednesday, or Thursday. In an eight hour day accuracy of judgement was the best during the fourth, fifth and sixth hours.

Since our appreciation of flavour and our judgement are infuenced by many psychological factors, the environment must be carefully controlled. However, some of the psychological factors that cannot be controlled are the experience background of each tester and his personal problems and reactions. Food experience affects everyone. If you have frequently eaten food highly seasoned with garlic, you will usually like that flavour but it will have nothing of newness or surprise for you. A person adventurous in other life experiences will find pleasure in a new eating experience; while the more conservative will dislike anything unfamiliar.

The *consumer acceptance* or *preference* has been used in recent years for some products. If a new food is introduced, only one sample is offered to a large consumer panel, but if a food is modified, then two samples are submitted and a preference requested. Occasionally more than two samples are submitted but usually this leads to confusion in the subjects and the results are doubtful.

4.8 Flavour Intensifier: Monosodium Glutamate

Large quantities of monosodium glutamate have been used in the Orient for years and in the United States for approximately 2 decades as a flavour intensifier. In China and Japan fermented soybean curd is added to many foods and soy sauce is a common condiment. Soy sauce is rich in monosodium glutamate and interest led eventually to the production of relatively pure monosodium glutamate. It is claimed by some authors that this compound has little or no flavour itself but intensifies the flavour of meats and vegetables through a rounding or blending effect. It is widely used in stews, canned meats, soups, chowders, etc. It does not have an effect on fruits or fruit juices or sweet spicy foods. Cairncross and Sjostrom reported that it suppresses the sharpness of onion, the rawness of many vegetables, and the earthiness of potatoes. They also reported that the effect is noticeable in the pH range 3.5 to 7.2, the range in which most foods are eaten. They found that fats, oils, and high viscosity in foods modify the influence of sodium glutamate considerably.

The mode of action of sodium glutamate and its effect on flavour are still in dispute. Much of the work carried out has been an investigation of the preference of tasters for a food with and without added monosodium glutamate, and the underlying physiology and biochemistry of the effect.

Glutamic acid occurs in high concentration in numerous proteins. Wheat gluten for example contains approximately 36 per cent. The free glutamic acid hydrochloride is prepared commercially from wheat, corn or sugar beet wastes by acid hydrolysis of the protein. Glutamic acid is formed by dissolving the crude hydrochloride, adjusting the pH to 3.2, and allowing it to crystallize slowly. The product is neutralized with either sodium hydroxide or sodium carbonate, decolourized and crystallized. The product is relatively pure monosodium glutamate.

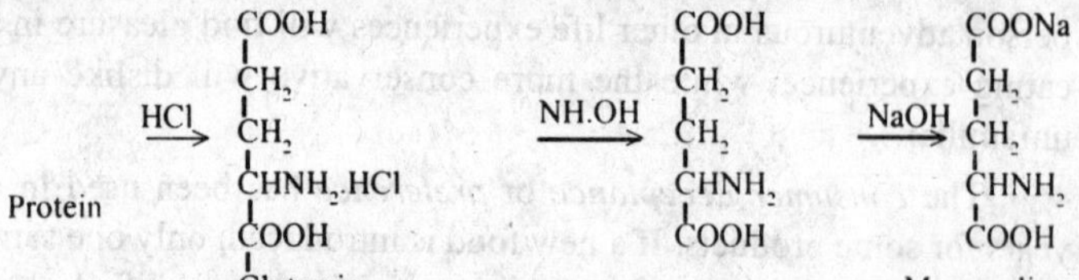

Glutamic Acid Hydrochloride

Monosodium Glutamate

Very large amounts of this material are now used commercially in the production of many meat and vegetable dishes. Although it has been claimed that it is not detectable and although it is true that pure monosodium glutamate is only slightly sweet and salty, the author has noticed that the presence of this "intensifier" at the level usually used in foods can be recognized. It gives a sameness of flavour to foods as dissimilar as chicken or tuna pies and makes the product markedly different from the freshly prepared dish. On this basis, it is highly objectionable to her.

4.9 Flavouring Extracts

These are widely used in the food industry, and in the United States are quite well standardized. The Food and Drug Administration defines an extract as "a solution in ethyl alcohol of proper strength of the sapid and odourous principles derived from an aromatic plant, or parts of the plant, with or without its colouring matter, conforming in name to the plant used in its preparation." If artificial colouring or synthetic flavouring compounds are added, it is required that the label state this. The flavouring extracts are made either by adding the essential oil to alcohol or to water and alcohol, or by percolating the chopped plant or plant part with a mixture of water and alcohol. Thus lemon extract is prepared by adding the lemon oil expressed mechanically from lemons, to alcohol or diluted alcohol. The Food and Drug Administration's standards require 5 per cent by volume of lemon oil. Also acceptable is an extract prepared after the terpenes, which add little to the flavour, have been removed.

Vanilla extract, the most widely used extract in the United States, is prepared by percolating macerated vanilla beans with alcohol, frequenlty in the presence of glycerol or sucrose. The flavourful compounds extracted are numerous but the most abundant is vanillin. Since vanillin can be prepared synthetically very cheaply, synthetic vanilla is sold in large quantities. Its flavour is not quite as fully rounded as real vanilla extract but it is used widely in many foods selling at relatively low cost. Usually coumarin is added to the vanillin in the synthetic extract to improve the flavour, and caramel or a food dye to stimulate the brown colour extracted from the vanilla beans. Discussion of the methods of preparing flavouring extracts can be found in a textbook of food technology.

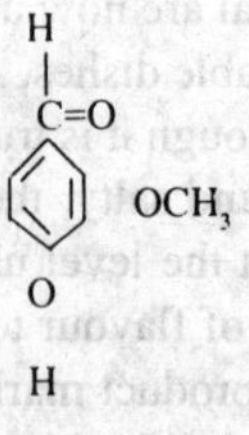

Vanillin

4.10 Synthetic Flavouring Substances

These have been produced for many years. Originally esters were used in small amounts for imitation flavours, particularly in cheap confections. Although an ester such as amyl acetate is reminiscent of banana flavour, simple esters are crude, rough substitutes for natural flavour. Analysis of natural flavours has developed slowly because of the difficulty of separating substances that occur in a concentration as low as that in most foods. New methods of study (see below) are now opening the way for careful and detailed analysis of the flavouring compounds in many foods. During the past 50 years numerous compounds have been synthesized which are useful in synthetic or imitation flavours. The common so-called *flavourmatics* are listed in a table prepared by the Food Additive Committee of the Flavouring Extract Manufacturers Association of the United States. (See Table 4.1.) These are arranged in order of their importance. The type of flavour in which they are usually used and the parts per million concentration in the finished food are given.

Table 4.2 shows the composition of a few imitation flavours. It is interesting because it shows the complex nature of some of these imitation flavours. The list of compounds isolated from natural flavours is far from complete. Many more compounds have now been identified in strawberry, apple and coffee flavours.

4.11 Recent Developments in Flavouring Research

Recent developments in flavouring research have followed three principal paths: (1) careful identification of the composition of some flavours, particularly through the use of gas chromatography as a means of separating the large number of compounds that occur in low concentration in many natural foods, (2) production of natural essences by condensation of the vapours formed in vacuum or low temperature concentration of foods, and (3)the search for flavour precursors and their enzymes. As the food industry moves farther along the path of prepared or partially prepared foods, the problem of flavour in the final product becomes more troublesome. With the concentration of populations in cities and the rise in standard of living, a growing demand for quality food occurs at the same time as a need for food products stable enough to allow time for distribution. Maintenance of flavour is one of the chief problems in producing quality food.

Table : 4.1 The Primary Flavourmatics

Aromatic Chemical	Flavour	Avg. Use (P.P.M.)	B.P.°C
1	2	3	4
Group 1			
Vanillin	vanilla	31.5	81.5° MP
Ethyl Vanillin	vanilla	16.6	76.5° MP
Citral	lemon	17.6	229°
Benzaldehyde	cherry, almond	84.8	180°
Cinnamic Aldehyde	cinnamon, cola	110.7	252°
Methyl Salicylate	root beer wintegreen	129.7	221°
Amyl Aceate	banana, fruit flavors	78.4	142°
Amyl Butyrate	banana, fruit flavors	23.7	185°
Ethyl Acetate	fruit, rum	92.7	76°
Ethyl Butyrate	strawberry, fruit	31.1	120°
Methyl Anthranilate	grape	97.1	255°
Ethyl Oxy Hydrate	rum	472.7	
Anethole	root beer, anise	133.5	235.3°
Safrol	root beer, anise	16.9	236°
Menthol	mint	111.2	215°
Group II			
Heliotropine	vanilla, cherry	3.4	263°
Aldehyde C-18	coconut	17.6	263°
Diacetyl	butter	17.3	88°
Aldehyde C-16	strawberry	10.0	271°
Ethyl Oenanthate	grape	8.1	187°
Allyl Caproate	pineapple	11.7	187°
Aldehyde C-14	peach	8.0	297°
Eugenol	spice	48.8	253°
Amyl Valerinate	peach, fruit	9.6	200°
Ethyl Valerinate	fruit	14.4	145°
Carvone	spice, mint	190.3	230°
Anis Aldehyde	cherry, vanilla	3.6	247°

(Contd.)

1	2	3	4
Group III			
Benzyl Acetate	strawberry, fruit	8.8	215°
Ethyl Formate	rum, strawberry	77.0	54°
Ionone	raspberry	0.9	147.5°
Ethyl Lactate	grape	42.1	154°
Tolyl Aldehyde	cherry	9.3	204°
Amyl Caproate	pineapple, apple	4.4	222.2°
Linalool	spice	30.2	199°
Citronellal	rose (general)	14.20	206°
Butyl Acetate	fruit	26.4	126.5°
Ethyl Pelargonate	grape, fruit	2.3	228°
Butyl Butyrate	fruit	17.9	166.4°
Aldehyde C-10	citrus	1.2	208°
Ethyl Heptoate	grape, pineapple	4.6	172.1°
Geranyl Acetate	fruit	8.4	245°
Methyl Phenyl Acetate	honey	2.4	222°
Group IV			
Aldehyde C-8	citrus	0.71	165°
Aldehyde C-9	citrus	2.81	191°
Cyclohexyl Butyrate	pineapple, fruit	8.3	212°
Ethyl Laurate	spice, fruit	6.9	269°
Linalyl Acetate	citrus	5.4	220°

Table 4.2 : Synthetics

	Aromatic Isolated	Sample Formula	%
1	2	3	4
Raspberry	beta ionone	beta ionone	10
	anisic aldehyde	iso-butyl acetate	23.49
	benzaldehyde	anisic aldehyde	1
	phenyl ethyl alcohol	phenyl ethyl alcohol	3
		phenyl ethyl iso-butyrate	4
	hexene	ethyl methyl p-totyl glycidate	35
	n-hexanol	phenyl ethyl anthranilate	7
	iso-amyl alcohol	vanillin	3
	iso-butyl alcohol	hexyl butyrate	0.5
		iso-amyl acetate	4
		ethyl n-butyrate	5
		rose oil	4
		diallyel sulfide	0.01
Strawberry	borneol	ethyl methyl phenyl glycidate	35
	methyl caproate	amyl aldehyde	0.5
	amyl aldehyde	bornyl acetate	0.5

(Contd.)

1	2	3	4
	ethyl caproate	ethyl caproate	2
		vanillin	3
		beta ionone	9
		ethyl methyl p-tolyl glycidate	12
		.iso-butyl acetate	30
		ethyl butyrate	8
Cherry	geranyl butyrate	cinnamyl anthranilate	3
		iso-amyl acetate	12
	.ethyl caprate	iso-butyl acetate	12
	terpenyl butyrate	cinnamic aldehyde dimethyl acetal	3
	ethyl caproate	benzyl alcohol	5.5
	benzaldehyde	geranyl butyrate	2
	geraniol	ethyl caprate	4
		terpenyl butyrate	0.5
		ethyl caprorate	2
		benzaldehyde	8
		p-tolyl aldehyde	3
		p-methyl benzyl acetate	11
		ethyl methyl p-tolyl glycidate	16
		vanillin	7
		heliotropin	5
Peach	linalyl esters (as valerate, acetate, formate)	linalyl formate	10
		gamma undeca lactone	13
		linalyl butyrate	4
	furfural	heliotropin	14.5
		geranyl valerate	15
		furfural	1.5
		alpha-methyl furyl acrolein	6
		methyl cyclo pentenolone valerate	10
		benzaldehyde	5
		iso-amyl formate	15
		iso-butyl butyrate	6
Apple	geraniol	geranyl valerate	10
	acetaldehyde	geranyl n-butyrate	8
		geranyl propionate	8
		linalyl formate	10
		iso-amyl valerate	15
		vanillin	8
		allyl caprylate	6
		geranyl aldehyde (citral)	5
		acetaldehyde	6.5
		methyl cyclo pentenolve valerate	8
		alpha methyl furyl acrosin	2
		iso-amyl butyrate	13.5

1	2	3	4
Coffee	α furfuryl mercaptan	α-furfuryl mercaptan	10
		ethyl vanillin	3
		solvent	87

The gas chromatograph is a fine instrument for the separation of the many compounds that make up the flavour complex in a natural food. A column is packed with finely divided particles of a solid, and wet with a liquid of low volatility. The nature of the liquid affects the separation achieved and as in all types of chromatography, the discovery of the suitable liquid is often one of the most time consuming parts of the research. A concentrate of the essence is made by one of a number of methods—for example, extraction or distillation—and this concentrate is placed on the column. The components are eluted by a gas such as helium that removes them from the column in fractions or as single compounds. The fractions are run on other columns and then separated into single compounds. Separations of very small quantities of material can be made very rapidly in a matter of minutes and the recoveries from the columns are almost quantitative. With large samples the gas chromatograph can be used as a preparative device.

After the components of a mixture are separated, the classical methods of identifying compounds can be used. In regular paper or column chromatography, the identity of a compound is often established by comparison of its behaviour with that of a known compound. This technique may be applied to gas chromatography. Relative amounts of substances are recorded.

The flavours of many foods are now being studied by this method and in a fairly short time the chemistry of flavour will be on a firm foundation for the first time. An understanding of changes in flavour on processing or aging of a food such as coffee is at last possible with this technique. (See Tables 4.3. and 4.4.)

Table 4.3 : Compounds Found in the Gaseous Emanation of Onions

Compound	*Relative Amount*
Propionaldehyde	very abundant
Methyl Alcohol	
Propyl Mercaptan	abundant
Hydrogen Sulfide	
Acetaldehyde	small
Sulfur Dioxide	
Dipropyldisulfide	
Propyl Alcohol	trace
4-Hexen-l-al	
2-Hydroxy Propanthiol	

During the processing of a number of foods, particularly fruit juices, the loss of aroma during concentrating is appreciable. Many years ago vacuum evaporation was introduced with low temperature condensers to trap as much of the volatile compounds as possible. This was a great improvement in many cases over evaporation at atmospheric pressure. However, since air is present in the original product, venting the system is necessary; consequently, enough volatile material is lost so that the fresh flavour disappears. Improvements in design now allow the removal of most aroma in the first fraction by flash evaporation. The juice is heated rapidly in a matter of a few seconds by passing it with turbulent flow through a small pipe and then it is rapidly cooled. Flash evaporation is carried out at atmospheric pressure if the volatiles will take the high temperature, otherwise at low pressure. The condensate from the flash evaporation passes to a fractionating column and is separated into two fractions. The vent gases are scrubbed by a counter-current of cold column bottoms that trap the most volatile compounds by dissolving them. The essence is stripped from the column.

Table 4.4 : Volatile Components in Coffee Aroma

Formic Acid	Phenol
Acetic Acid	Resorcind
Methyl Ethyl Acetic Acid	Cresols
n-Valeric Acid	Ammonia
iso-Valeric Acids	Methyl Amine
Higher Fatty Acids	Trimethylamine
Ethyl Alcohol	Pyrrole
Acetyl Methyl Carbinol	n-Methyl Pyrrole
Furfuryl Alcohol	Pyridine and Homologues
Acetaldehyde	Pyrazine
Methyl Ethyl Acetaldehyde	Guaiacol
Furfural	p-Vinyl Guaiacol
Acetone	Eugenol
Diethyl Ketone	Hydrogen Sulfide
Diacetyl	Methyl Mercaptan
Acetyl Propionyl	Dimethyl Sulfide
Hydroquinone	Furfuryl Mercaptan
Esters	Furane
Furfuryl Formate	Sylvestrine
Furfuryl Acetate	Vanillone
Methyl Alcohol	n-Heptacosane
2,3-Dioxyacetophenone	

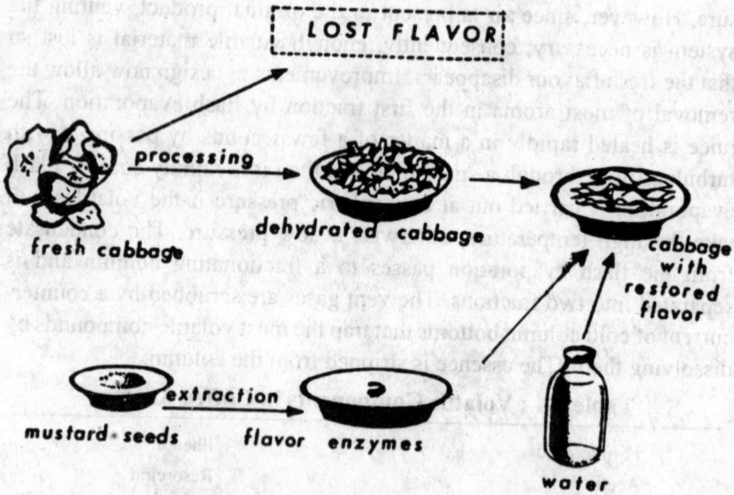

Fig. 4.1: Restoration of Natural Flavour to Dehydrated Cabbage.

With the progress of biochemistry and the rapid rise in our understanding of the changes that occur in plant and animal cells, the time is ripe for some investigations of flavour precursors and attempts to utilize biochemical reactions for the maintenance and enhancement of fresh flavour. Studies are underway to find methods for the isolation or at least the concentration of flavour precursors and enzyme systems capable of developing fresh flavour. It is hoped that when these are obtained their addition to dried or concentrated foods will improve the flavour and acceptability even in the presence of off-flavours.

5

Germs in Food Items

5.1 Introduction

The presence of various types of germs in our food items are one of the main problems that need our utmost attention in discussion of any food hygiene and is a food hazard. It is for this reason that we must know about the germs before we proceed further to discuss the topic.

A germ can produce many thousands of millions of off springs with in 24 hours if the conditions are favourable to it, however if the conditions are against it, a germ may die in a few seconds.

To grow germs the scientists have to prepare special media. Such a media is generally prepared by using jelly or broth and mixing with it chosen carbohydrates, proteins, vitamins etc. The pH is suitably adjusted. To this medium is then added a piece of the substance thought to be contaminated mixed with sterile water or saline is added to the medium. The bottle of broth or plate of jelly is placed in an incubator which must be kept at a set temperature; if this is only a degree or two wrong, the germ may not grow. Even the amount of oxygen must be regulated; some germs require a lot, some cannot stand more than a trace.

Some germs will not respond to all this care. They grow only on living cells, and the bacteriologist must get such cells from animal or human tissues and keep them alive in the laboratory so that he can try to grow these germs on them. These are called tissue cultures and are used mainly for growing viruses. Viruses are very small germs which grow only in living cells. However, some germs will not grow even in tissue cultures, but only if injected into live animals. This is why animals like rats, mice and guinea-pigs have to be used in laboratories.

5.2 Laboratory Life

The germs are all over the place and they do not have much difficulty in surviving outside the laboratory. They can thrive on the skin of one's nose, on the wing of a dead bird or in the dry dust on a window ledge. This, is true and it is the artificiality of the conditions in the laboratory that makes things so difficult. It is very much the same with human beings. In the artificial conditions which have to be set up round a diseased or injured patient, keeping the patient alive can be an extremely difficult affair. He may require a machine to keep his breathing going, the amount of oxygen he requires must be calculated, and measures must be taken to regulate his temperature. If he cannot take food by mouth, nourishment must be provided directly into a vein, and this can raise many problems. It is easy to give a carbohydrate like glucose intravenously, but very difficult to give even the simplest proteins or fat. The amount of fluid that goes into and comes out of the patient's body must be measured. A highly skilled team of doctors and nurses is needed. A visitor from outer space seeing this for the first time might conclude that human beings led a most precarious existence. He would not realize that the same human being, when well, gives not a thought to how he breathes or how his temperature is maintained, drinks as much as he likes, and eats anything that takes his fancy from rhubarb to sharks' fins or birds' nests. Neither man nor germs show up well under artificial conditions.

Under natural conditions germs flourish almost anywhere. Some food-poisoning germs can live in garden soil for as long as a year. They have also been found growing in waste-pipes from wash-basins and water-closets, and in dry floor dust. They can live in water for weeks on end, but they can also survive in a dry crack in a table top; but they are happiest living on some other animal as parasites. Staphylococci, for example, live very well on human skin. They may be found in the fold of a nail, on the edge of an eyelid, deep in the armpit or anywhere else on the body. Normally they do us no harm, but they can cause a whitlow on the finger, a stye on the eye, an abscess in the armpit or some much more serious condition. From our skin they easily get on to towels, dish cloths, rolling-pins and table napkins. If they get into our food, they can cause food poisoning.

Fig.5.1 : Germs flourish almost anywhere: the arrows point to some of the places.

Salmonellae live in the intestinal tract of many animals: the list includes rats, pythons, tortoises, pigeons, cattle and pigs as well as man. They get into the host by first multiplying in its food supply and then being swallowed. All kinds of food can be invaded by food-poisoning germs, though different germs prefer different foodstuffs. However, germs cannot jump on to food. They must be placed on it or carried to it, and so it is usually food that is handled a lot that becomes contaminated with germs. Pies, cold meats, salads, trifles and creams are obvious examples. These are all foods that may be prepared a long time before being eaten. This is another important point; one or two germs on food can easily be eaten without harm, but if the germs have had time to multiply in the food, the one or two become many millions and an outbreak of food poisoning results.

5.3 Geography of Germs

Germs are found all over the world, and nowadays they do not stay put, for modern methods of food production and distribution give them every chance to roam. Germs have, in fact, become great travellers, and round-the-world trips are commonplace with them. Heidelberg is a pleasant German university town, but it is also the name of a food-poisoning germ which was first described there in 1933, but has since become common all over Europe and in America. Johannesburg, Montevideo, Tennessee and Singapore, among nearly a thousand other places, have also given their names to germs that have since wandered far from home. This ability of germs to survive long periods of travel is one of the characteristics of food-poisoning germs to which we shall have to refer frequently in later chapters.

5.4 How Germs Survive

Germs have a fairly tenacious hold on life. They can put up with all sorts of discomfort. Most of them can survive drying. Dysentery germs, for example, can live for weeks on bits of blotting-paper, on cotton-wool balls, on bits of dry sticks and on metal tags. They have survived for over a fortnight on the wooden seat of a water-closet, a fact which raises an obviously important point of personal hygiene. This resistance of drying is true of many of the food-poisoning germs, so that if they once get into a kitchen, a bakehouse, or any place where food is handled, they can persist in the dust on tables, floors and ledges, and be a source of danger for quite sometime.

(i) Effect of Cold

Germs can stand almost any degree of cold. It is in fact customary to preserve germs in laboratories, especially viruses, by storing them at — 70°C. Freezing-point is 0°C., so that — 70°C. is cold indeed, but some germs can stand such temperatures for months or years on end. Germs can therefore easily survive in frozen food. They cannot multiply at such temperatures. In fact, they stop multiplying at temperatures well above freezing-point. That is why we keep food in refrigerators to prevent germs from multiplying. If there are one or two germs on the food when it is put into the refrigerator at night, there will still be one or two on it in the morning. If the same food had been left in a warm room, there might have been countless millions of germs in the food by daybreak.

(ii) Effect of Heat

Germs like a little warmth, but most of them cannot stand heat, and are very easily destroyed by it. Most salmonellae, for example, are killed in 15 to 20 minutes at 60°C. (140°F.); this is 40° below boiling-point and is not really very hot. Most cooking methods heat the food well above 60°C. This is lucky for us, though unlucky for salmonellae. Some food-poisoning germs are not so easily destroyed by heat. *Clostridium welchii* spores survive boiling for several hours, so that if the germ gets into food and multiplies in it, cooking may not destroy it. Staphylococci are easily killed by cooking, but if they have already produced toxin in the food, the toxin may resist the heat of cooking though the germ itself dies.

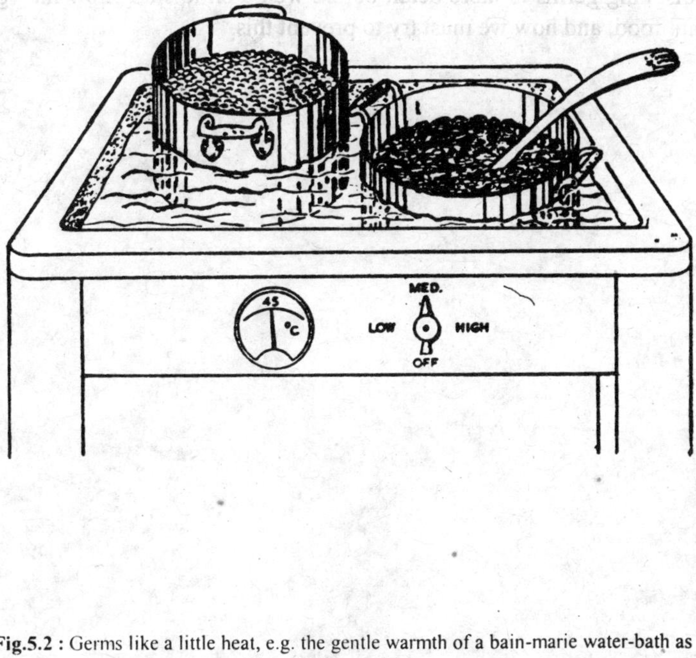

Fig.5.2 : Germs like a little heat, e.g. the gentle warmth of a bain-marie water-bath as used in catering establishments.

5.5 Spores

Some germs exist in two different forms—the vegetative and the spore form. These can best be explained by thinking of the life-cycles of some animals. A squirrel is full of life in the summer, but in the winter it hibernates, or goes into a state where all its living processes are slowed down; it can sleep in this state for weeks on end without waking for food. The spore of a germ can be likened to the hibernating state of the squirrel, the vegetative form to its active summer life. In the sporing state, a germ can survive for many years. Spores can often resist very high temperatures. The vegetative form of the germ which causes botulism, *Clostridium botulinum*, is killed by heating for 10 to 15 minutes at 80°C. (176°F.), but its spore can resist boiling for hours. This is obviously important if a spore gets into food, but not all germs form spores. A germ in the vegetative state is full of life and multiplies quickly. In this state it is usually easy to kill by heat and other means.

In this chapter we have seen in general terms how germs live and die. In the next chapter we shall consider the life histories of food-poisoning germs in more detail before we go on to study how they get into food, and how we must try to prevent this.

6

Food-poisoning Germs

6.1 Introduction

The gastro-intestinal gets upset if any germ gets into food and multiplies quickly in it. Even those germs that are normally harmless can cause this, but only when the food is prepared under very bad conditions with no thought for food hygiene. The germ that is present in everyone's intestinal tract, *Escherichia coli*, is one such germ. It normally causes no harm so long as it stays in the bowel. Actually it probably carried out important functions in the bowel; but it sometimes causes trouble when it gets into food. There are, however, germs that have a special liking for human food; they multiply very readily in it, and cause typical symptoms in people who eat the contaminated food. These include the *salmonellae, some clostridia, shigellae* and *staphylococci*. These will be considered in some detail.

6.2 The Naming of Germs

Germs get their names in various ways. *Escherichia coli*, is called after the bacteriologist Escherich who first described it in 1885. Colon is the Greek word for bowel, so that *Escherichia coli* is simply a bowel germ called after its discoverer. Mercifully it is usually called just *Escherichia coli*. Salmonellae are called after Dr. Salmon, and Shigellae after Dr. Shiga. Shigellae are the germs that cause dysentery. Dysentery itself comes from another Greek word, *enteron* meaning bowel, and the *dys* signifies something wrong. A well-known dysentery germ is *Shigella sonnei*, and Dr. Sonne was a Danish doctor who described it. Staphylococci comes from two Greek words, *staphylee*, that means a bunch of grapes, and *coccus* that means berry, and these germs under the microscope look like a bunch of grapes or berries. *Clostridium welchii* is a germ called after Dr. Welch, but in *Clostridium botulinum* the second word comes from the Latin, *botulus*, meaning sausage. The term was first used after an outbreak of botulism in Germany in 1793 which caused severe illness in 13 people who shared a large sausage. Clostridium comes from a Greek word, *closteer*, meaning a thread or spindle, and refers to the shape of the germ when swollen by its spore. Brucellae are the cause of brucellosis, a milk-borne disease, and the germ is named after Dr. Bruce who found this germ in the spleen of four patients who died of the disease in Malta in 1887.

6.3 Families of Germs

Germs, like human beings, belong to specific families; they have a family name and an individual name. The family name always comes first and the individual name second. *Salmonella dublin* is the full name of a germ in the salmonella family; its individual name dublin comes from the town Dublin where it was first described. The family name always have a capital when the full name is used, and the second name a small letter, even when it is the name of a person or town; and the full name is always printed in italics. Other places which have given their names to salmonellae are Aberdeen, Brandenburg, Chicago, Georgia, Kentucky, London, Newport and Singapore, and there are many more. Some salmonellae are named after their discoverer, e.g. *Salmonella thompson* and *Salmonella virchow*; some after the animals they infect, e.g. *Salmonella pullorum* and *Salmonella gallinarum*, both parasites of poultry—the second names are both Latin for hens. Some are called after the diseases they cause, e.g. *Salmonella typhi*, which causes typhoid fever. The first name may be contracted into Sal. or just S. One of the commonest of all is *Salmonellae typhi-murium*; the name suggests that it causes typhoid in mice (*mus*), though this is not true. Another fairly common name is *Salmonellae cholerae-suis*; it causes a diarrhoeal disease in the pig (*sus*) but it is not correct to call this cholera, because the disease is caused by a germ of different family.

All this may appear to have nothing to do with food poisoning, but when one is trying to get an exact picture of any problem, as we are of food poisoning, it is as well to use the terms exactly, and there is no harm is knowing what words mean.

6.4 Salmonellae

Salmonellae are short, thin, rod-like germs. They propel themselves along with the of fine threads or flagella that stream out from the sides of the rod and are in constant motion. Under the microscope they appear as very badly trained rowing eight, except that there are far more than eight pairs of oars of flagella, and the course steered is a very irregular one. The word flagella means whips, and flagella look more like twisting whips than rigid oars.

Individual salmonellae can be seen only under the microscope, but when they are grown on agar jelly they form colonies consisting of thousands of germs and these colonies are two to three millimeters in diameter and can be easily seen on the surface of the jelly. The salmonella family of germs is a very big one with over 1,000 different members. These salmonellae, although they share the main family features, all differ slightly from one another, and these differences can be detected by very sensitive laboratory tests. They are all given distinctive names as—*S. typhi-murium, S. dublin, S. heidelberg* etc.

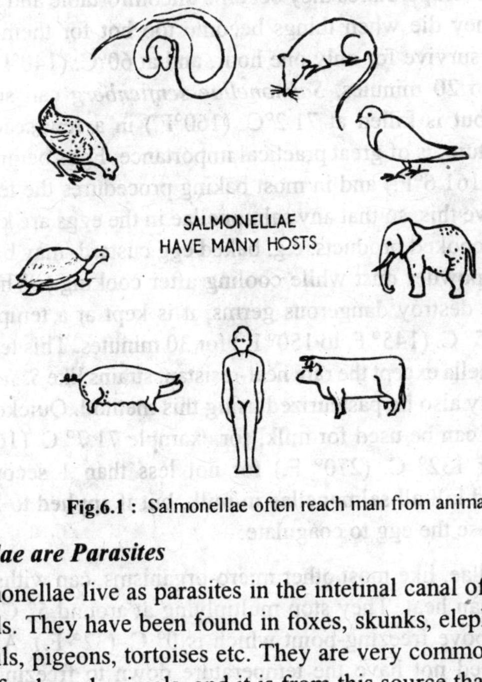

Fig.6.1 : Salmonellae often reach man from animals.

Salmonellae are Parasites

Salmonellae live as parasites in the intetinal canal of many different animals. They have been found in foxes, skunks, elephants, snakes, seals, quails, pigeons, tortoises etc. They are very common parasites of farmyard fowls and animals, and it is from this source that they usually spread to man.

Salmonellae can survive in soil for a year, and for long periods in dust on floors and tables. They have been grown from the waste pipes of water-closets and of wash-hand-basins, particularly in bakeries and factories where contaminated food may be handled. They can be grown from swabs dipped in sewers into which the feces of infected animals or humans is passed, and sewer swabs are sometimes used to trace the path of infection in outbreaks of disease. The ultimate source of salmonellae is always the intestinal canal of some animal, because it is only there that they can flourish and multiply: but if they get into man's foodstuffs and are given the right conditions of temperature and moisture, they find things very much to their liking there too.

Effect of Heat and Cold

Temperature is quite important. Salmonellae like the same body temperature as man, 37°C. (98.6°F.). This is one reason why they like a parasitic existence—it gives them a nice, equable climate in which to live. At higher temperatures they become uncomfortable and stop multiplying, and they die when things become too hot for them. At 55°C. (131°F.) they survive for only one hour, and at 60°C. (140°F.) they are killed in 15 to 20 minutes. *Salmonellae senftenberg* can survive this temperature, but is killed at 71.2°C. (160°F.) in a few seconds. This latter temperature, is of great practical importance. Eggs begin to coagulate at 72°C. (161.6°F.) and in most baking procedures the temperature rises well above this, so that any salmonellae in the eggs are killed. (The danger is that cooked products, e.g. baked egg custard, may be contaminated by egg powder dust while cooling after cooking.) When milk is pasteurized to destroy dangerous germs, it is kept at a temperature of 62.6°C. to 65.5° C. (145° F. to 150° F.) for 30 minutes. This temperature kills all salmonella except the rare heat-resistant strains like *S. senftenberg*. Liquid egg may also be pasteurized using this method. Quicker pasteurizing methods can be used for milk, for example 71.7° C. (161° F.) for 15 seconds or 132° C. (270° F.) for not less than 1 second. These methods would kill all salmonellae in milk, but if applied to liquid egg they might cause the egg to coagulate.

Salmonellae, like most other micro-organisms, can withstand cold much better than heat. They stop multiplying at around 5° C. (41° F.). This is well above freezing-point which is 0° C. (32° F.). A domestic refrigerator need not have the temperature down to freezing-point to stop germs multiplying—a temperature around 5° C. (41° F.) is adequate. When a deep-freeze refrigerator is used to freeze and preserve food, the temperature falls far below freezing-point, but salmonellae in the food, for example inside a chicken, are not killed by this treatment. When the bird is thawed out and the temperature rises above 5° C.

(41° F.) they begin to multiply again. The point to remember is that *refrigeration does not kill germs.*

Effect of Moisture

Salmonellae, like other living things, require water. They can, however, survive for long periods in dry dust or in powder, but they cannot multiply in the absence of moisture. Salmonellae are generally present in untreated dried-egg powder and they have occasionally been grown from dried milk. As soon as we add water to the powder and the temperature is right, the salmonellae start multiplying again till soon the liquid egg or the reconstituted milk is teeming with germs.

Effects of Chemicals

Ordinary disinfectants like the phenols, etc. kill salmonellae quickly; these cannot be used on foodstuffs, but chlorine and potassium permanganate can be used. Lettuce can be rendered germ-free if immersed in water containing 80 parts per million of chlorine for 30 seconds; chicken carcasses need a concentration of 200 parts per million for 10 minutes. Antibiotics may also kill salmonellae if added to washing fluids or the slush ice used in poultry plants. The snag is that some salmonellae, resistant to antibiotics, may make their appearance; these can then flourish in spite of the antibiotic and lead to a very false sense of security.

6.5 Shigellae

These are the germs that cause dysentery, they are short, rod-like germs. They form tiny colonies on special agar jelly that contains several sugars, including lactose. Most intestinal germs attack lactose and form acid from it and when this occurs an indicator in the jelly turns the colonies pink. Shigellae do not attack lactose and so their colonies remain white and can be easily distinguished from the other germs on the plate. They are called non-lactose fermenters.

The shigellae family is a small one and it consists of just over 30 members—the salmonella family has more than 1,000 members. Three well-known members of the shigella family are *Shigella sonnei, Shigella flexneri* and *Shigella shigae,* all called after the doctors who first described them.

Shigellae live as *parasites* of the intestinal canal of man, and man is their only host: this is very different from the salmonellae which have very many hosts. If an outbreak of food-borne dysentery takes place, there must be somewhere in the chain of infection a human carrier of the germ, most probably in the kitchen.

Shigellae can also exist outside the human body. They often survive for long periods in dust, on tables, on cloths and even in sand. They like a warm, damp atmosphere in subdued light, and may linger for days in and around water-closets. Shigellae can live on the fingers for several hours and can pass from hand to hand very easily, and then from hand to mouth. This is the way dysentery mostly spreads, from person to person, but if the germs do get into food, they multiply in it and outbreaks of dysentery occurs.

Effect of Heat and Cold

Shigellae survive for long periods at room temperature and at body temperature, but they are readily destroyed by moderate heat, almost exactly the same as are Salmonellae. They survive freezing for weeks together, so they are not killed by being placed in a refrigerator or a deep-freeze.

Effect of Moisture

Like all germs, shigellae needs moisture for multiplication, though they can get along in dry dust or on a piece of cloth for quite sometime. They prefer conditions of high humidity, that is where there is lot of moisture in the air.

Effect of Chemicals

Shigellae are quite easily killed by disinfectants. Phenol 1 per cent kills them in 15 minutes, and a solution of benzalkonium chloride 1 per cent in one minute; this latter can be used as a hand-washing solution during outbreaks of dysentery.

6.6 Staphylococci

Staphylococci are rather stout, round germs that usually cluster together like bunches of grapes — *staphylococcus* means a bunch of grapes. They have no flagella or whips and so are non-motile. A single staphylococcus is invisible to the human eye, but can be seen easily under a low-power microscope lens. They form colonies readily on agar jelly and they look like tiny pieces of butter scattered on the surface of the jelly. Some of the colonies are coloured—white, yellow or gold—and, to some extent, this colour helps to identify the more dangerous germs. *Staphylococcus albus*, the white staphylococcus, is generally harmless, but golden, *Staphylococcus aureus*, is dangerous.

Staphylococci form a family of several hundred members, and it is important to be able to distinguish them. This can be done by exposing a staphylococcus to even smaller micro-organisms known as phages. The word phage comes from the Greek word, *phagein*, meaning to eat, and these phages actually attack and destroy staphylococci. However, they are very 'choosey': one phage will attack only a few members of the staphylococcal family. Each phage is given a number, and a staphylococcus is referred to by the numbers of the phages that attack it. Thus staphylococcus phage type 6/47 is attacked by phages 6 and 47; staphylococcus 29/52/6/7/42E/47/54/73/75 is attacked by all the phages represented by these numbers. This typing is a piece of very skilled bacteriological technique, and in the investigation of an outbreak of food poisoning, such typing is essential to ensure that no false trails are followed. It is quite common to find a staphylococcus on the hands or in the nose of a food handler, but one must know its phage type before one can say it is the same staphylococcus as the one that caused the outbreak.

Not all staphylococci can cause food poisoning, but only those which produce an intestinal toxin known as enterotoxin. It is this toxin that causes the vomiting and diarrhoea of food poisoning, not the staphylococci themselves.

Staphylococci are parasites, and their commonest host is man. Over 50 per cent of normal people carry staphylococci in their nose and throat, and half of them have staphylococci on their hands also. Not all these germs form enterotoxin, but half of them do, so that at least 20 per cent of normal people carry food-poisoning staphylococci in their noses or on their hands. Thus if five people are employed in a kitchen, at least one of them is likely to be carrier of these dangerous staphylococci. This does not mean that an outbreak of food poisoning is bound to follow. The germs have still got to get into the right type of food and be given the right conditions of temperature and moisture in order to multiply. It is only when multiplying that staphylococci produce enterotoxin. Food hygiene consists in preventing germs getting on to food, or of multiplying if they do.

Staphylococci, although essentially parasites of men and animals, can exist outside the body and may be grown from towels, tables, knives, dishes or anything that a carrier touches. They may be grown from the air or dust in a contaminated kitchen or bake-house. Flies in such an environment readily become infected and may carry the germs on their feet, on their mouthparts, or in their intestines and infect food if they sit on it.

Staphylococci are present in cowsheds and can easily get into milk. They sometimes cause an inflammation of the udder in cattle, and then the milk may be teeming with staphylococci. Sometimes the staphylococci are the food-poisoning type and, as the germs multiply rapidly in milk, anyone drinking the milk raw may get food poisoning. Pasteurization of milk kills staphylococci so that pasteurized milk is safe milk.

Effect of Heat and Cold

Staphylococci get killed by heating to 60° C. (140° F.) for 30 minutes. This is well below cooking temperatures so that staphylococci will not survive cooking, provided the heat penetrates the food: this is not always the case inside a pie or large joint. Staphylococci grow rapidly at warm room-temperature, particularly in food, and, as they multiply, they produce enterotoxin. This toxin withstands boiling for 30 minutes, so that even if contaminated food is reheated before being eaten, the toxin is probably not destroyed and food poisoning may follow. The food in such a case is sterile, because all the staphylococci are dead, but the toxin is still there. This is why, in some outbreaks of staphylococcal food poisoning, staphylococci may not be found in specimens taken from patients: dead germs cannot come to life again in the laboratory.

Staphylococci multiply rapidly in warm milk; after a few hours standing at around 35°C. (95° F.) milk may be teeming with them: but, under modern conditions, milk is cooled quickly after being drawn and staphylococci have no chance to grow. In the early stages of cheese-making, staphylococci find temperatures much to their liking and they can outgrow the germs that are used to 'start' the cheese: but if cheese is made from pasteurized milk, as in all modern cheese factories, there is no danger, for any staphylococci present are killed.

Staphylococci are not killed by freezing: they may survive for months in the frozen state, though they cannot multiply. This is important in the making of ice-cream: if staphylococci have been allowed to grow and produce toxin when the mix was warm, the freezing will not make the ice-cream safe to eat.

Effect of Moisture

Staphylococci cannot multiply in the absence of moisture, however, they can stay alive for months in the dried state. They might for example survive for months in dried-egg or milk powder, and when the powder is reconstituted by the addition of water the staphylococci could begin to multiply again and produce enterotoxin.

Effects of Chemicals

Salts like nitrates and nitrites used in curing hams and other foods do not kill staphylococci: nor does pickling, for staphylococci can live in 10 per cent solution of salt. Antibiotics do kill staphylococci, particularly penicillin, but staphylococci often become resistant to them. Ordinary disinfectants like phenol are not very effective against staphylococci, which, however, are readily killed by the aniline dye disinfectants.

6.7 Clostridium Welchii

Clostridium welchii is one of the spore-bearing germs. It is rod-shaped with parallel sides, but when conditions prevent it from multiplying, the rod becomes swollen just short of the end. This swelling is the spore.

Clostridium welchii is primarily a native of the soil, probably it gets on to vegetables and other foodstuffs and is swallowed by man and animals. It is a common inhabitant of the intestinal tract. About 5 per cent of normal persons have the germ in their bowels and excrete it in their feces: in residential; institutions, with communal cooking, the percentage may rise to 20 or 30. Nearly 20 per cent of pigs harbour the germ, and the same percentage of rats and mice. It is not common in cattle, only in less than 2 per cent. The germ may be found in dust, water, milk and sewage. In fact, it is found almost anywhere, in the dust and air of kitchens, bakeries and shops. Flies and bluebottles are quite heavily infected with *Clostridium welchii* and so spread it to food. Swabs of the surface of raw meat are positive in 25 per cent of cases.

This germ does not grow willingly in the presence of air; it likes a situation where the amount of oxygen is reduced. It is, therefore, an anaerobic germ. It does not multiply on the surface of raw meat, but if it can find its way into the depths of a luke-warm stew, it finds ideal conditions of warmth and low oxygen, and so it multiplies rapidly and produces toxin. To produce food poisoning the food must contain live clostridia and the toxin, not the toxin only.

Effect of Heat and Cold

The non-sporing or vegetative form of the germ gets readily killed by ordinary cooking temperatures. Spores can resist boiling for hours on end. Well-cooked meat still contain live spores. As the meat cools, the spores germinate and produce the vegetative germs. These begin to multiply round about 50° C. (122° F.) and, if the meat is allowed to cool slowly, it may be heavily infected with clostridia by the time the temperature reaches 20° C. (68° F.). All this time the multiplying germs are producing toxin. If the contaminated food is now put in a refrigerator, multiplication stops, but the germ remains alive and the toxin remains toxic. Toxin is not destroyed by boiling, and in case the meat is reheated next day before being eaten, it will still be full of active toxin, and food poisoning may occur.

When meat is cooked some hours before being served, it is generally kept on a hot plate or in a *bain marie* till it is eaten. Such conditions are ideal for the multiplication of *Clostridium welchii*. If food is not to be served immediately after cooking, it must either (a) be cooked rapidly, put in a refrigerator overnight and be quickly reheated next day or (b) it must be kept at a temperature well above 60° C. till it is served. Otherwise there is a danger of food poisoning.

Effect of Moisture

The germs can grow only in the presence of moisture. Spores can survive dryness for very long periods.

Effect of Chemicals

The vegetative form of *Clostridium welchii* gets killed readily by ordinary disinfectants and by penicillin. The spores are very resistant to disinfectants and to penicillin.

6.8 Clostridium Botulinum

This germ is also a spore-bearer. There are at least 6 main members of the *Clostridium botulinum* family, lettered A to F. Most illness in man are caused by types A, B and E. The toxins produced by each type differ slightly from one another, but each toxin has a very similar and very powerful effect on the human body. The illness is very different from other forms of food poisoning and is often fatal.

Clostridium botulinum lives only in soil or on decaying vegetation, not in living animals. It is a saprophyte, not a parasite: *sapro* comes from the Greek word, *sapros*, meaning rottenness, and *phyte* means 'living on'. It is a very strict anaerobe and will not grow in ordinary atmospheric air. It may be present on vegetables and other fresh food, but does not multiply there and so does not produce toxin. Such food is therefore harmless. When food is preserved by pickling, or smoking or canning, the quantity of oxygen is reduced, and if *Clostridium botulinum* survived the preserving process, it could begin to multiply and produce the toxin in the food. If this food were then eaten uncooked, fatal botulism would follow, for the toxin is exceedingly powerful, and even a taste can kill. Fortunately *Clostridium botulinum* is nearly always killed in commercial preservation, but it can survive faulty home-canning.

Effect of Heat and Cold

The three things that must be considered are: (1) the vegetative or multiplying germs, (2) the spore, and (3) the toxin. The vegetative form of the germ is killed at 80° C. (176° F.), a higher temperature than is required for salmonellae or staphylococci, but still much below boiling-point and the temperature reached in canning. Spores are very resistant to heat. Type A spores resist boiling for hours, but are killed in 5 minutes at 121° C. (250° F.): the other types (B to F) are slightly easier to kill. Commerical methods of canning are adjusted to these temperatures: steam under pressure is needed. Botulinum toxin is destroyed quite easily by heat: all forms are killed in 2 minutes at temperatures below boiling-point. The toxins of types A, B and E (those that cause illness in man) are killed in 2 minutes at between 60° C. and 70° C. (140° F. to 158° F.), well below boiling point: types C and D require 80° C. (176° F.), while type F is destroyed at temperatures not much over 37° C. (98.6° F.). Thus if food is preserved at the correct high temperatures, spores get killed and the food is safe: but even if some spores survive, multiply in food and form toxin, this toxin would be destroyed quickly if the food were heated above 80° C. (176° F.), and the food could then be eaten without causing botulism. However, if the food were eaten without being heated, then all those who ate it are likely to get the disease, and many of them would die. Clostridium botulinum is a common germ of the soil, but cases of botulism are very rare.

The germs survive freezing. The vegetative form stops multiplying under 10° C. (50° F.) but begins to multiply again above that temperature. Spores and toxin survive freezing indefinitely, and any spores or toxin present in food when it is put into deep-freeze, will still be present when it is taken out.

Effect of Moisture

Clostridium botulinum requires moisture for growth, but spores survive for years in dry dust.

Effect of Chemicals

The acidity of food has a marked effect on the survival of *Clostridium botulinum.* At pH 7.0, spores resist boiling for hours: At around pH 5.0, the same spores are killed by boiling in 45 minutes, while around pH 3.5, boiling kills them in 10 minutes. Vinegar in a concentration of 2 per cent destroys the spores as also salt at 8 to 10 per cent.

This explains why the germ is killed by some preserving methods in which the temperature is too low. Tinned fruit, for example, is an acid product, and botulism has never been caused by eating it. String beans, seem to favour the growth of the germ, and botulism has been caused by eating home-canned beans. All the factors affecting the growth and destruction of *Clostridium botulinum* are not fully understood and constant watch must be kept on changing methods and changing fashions in food production and consumption, lest some minor alteration in the environment around the germ lead to its unexpected survival in the food.

7

Food Poisoning

7.1 *Introduction*

The term 'food poisoning' has an alarming ring about it, however, food poisoning is simply a form of entereities conveyed by food, and there is no need to be unduly anxious. The term 'food poisoning' is a useful one, because it emphasizes the fact that food is to blame. Caterers, cooks and all who handle food should certainly be worried and alarmed by outbreaks of food poisoning, for it is a condition which is preventable and therefore should never occur.

7.2 *Incubation Period*

It is the time that passes between the entry of the germ or toxin into the body and the occurrence of the first symptom. In staphylococcal cases, the toxin is present in the food and it irritates the stomach very quickly: the incubation period is short, generally from ½ hour to 4 or 5 hours. In salmonella poisoning there is no toxin: the germs need time to multiply in the patient's intestinal tract and cause symptoms only when they reach very high numbers: the incubation period is longer, from 12 to 48 hours. In *Clostridium welchii* poisoning, toxin is present but the germs also multiply inside the patient: the incubation period is between 12 to 24 hours. None of these figures is exact. A good deal depends on how heavily infected the food is: if there are very large numbers of salmonellae, for example, the incubation period might be only 8 to 12 hours, but a very short incubation period generally means staphylococcal poisoning.

7.3 *Symptoms*

The three symptoms common to all three forms of food poisoning i.e. diarrhoea, vomiting and pain. The incidence in the three forms of food poisoning is different.

Salmonella food poisoning. Diarrhoea is the main symptom. It varies in severity. In most patients it is moderately severe, with diarrhoea occurring every hour or so at the start but usually ceasing after the first day. In some it is much more severe and causes collapse, while in others it amounts to nothing more than a few loose stools which the patient hardly notices. At the onset the patient may have a little griping, abdominal pain but this is never severe. Vomiting is uncommon except when the food has been very heavily contaminated. The whole illness is usually over within two days, though the patient may continue to pass salmonellae in his faeces for much longer.

Sometimes salmonellae get through into the patient's blood-stream and the illness is then much more severe and needs specialized treatment. Such a patient can remain ill for several weeks, and the germ may settle down in some part of the body, particularly the bone marrow, and cause internal abscesses many years later. This invasion of the blood-stream depends partly on the quantity of infection in the food, but much more on the type of salmonella. *S. Choleral-suis* and *S. Virchow* are two types which can cause this type of illness.

Staphylococcal food poisoning. The most prominent symptom is vomiting. This is due to the toxin irritating the lining of the stomach. It is sometimes very severe and can lead to very great weakness or collapse. The vomiting helps to get rid of some of the toxin and this helps to cut the illness short. Vomiting generally stops within 5 to 6 hours but the patient may feel exhausted for the next 12 hours. Pain is not common, other than that caused by the vomiting. Diarrhoea is generally not severe and many patients have none at all.

Clostridium welchii food poisoning. In it pain in the abdomen is quite common. It can be fairly severe at first, a nagging, colicky pain in the upper abdomen and may last for several hours. It is followed by mild diarrhoea in nearly all patients. Vomiting is uncommon. This illness is generally over in one or two days.

The three types of illness differ a little in their symptoms. If most of the patients in an outbreak complain mainly of abdominal pain, the cause is probably *Clostridium welchii.* If vomiting is their main symptom, the cause is probably staphylococcal. If diarrhoea, with little pain or vomiting, is prominent, the outbreak may be due to salmonellae.

In all three forms, the illness is, on the whole sharp but short. In a few patients, especially frail old patients or patients already ill with another disease, it can be very severe or even fatal.

7.4 Laboratory Diagnosis

An exact diagnosis can be made only by bacteriological tests. Specimens of vomit, faeces and blood from the patient and any portions of the food should be sent to the laboratory. These specimens will be put into special media, and the tests done to identify the type of the germs. All this takes time and it may be several days or more than a week. A preliminary report can be given in 24 hours and this usually gives an accurate enough clue to those who are investigating the outbreak. When the outbreak is a staphylococcal one, the result are often negative: this is because the staphylococci themselves have been killed in cooking, and there is no satisfactory way of testing for the toxin in the laboratory.

7.5 Botulism

Botulism is a disease and its symptoms are very different from those of food poisoning. Its main effect is to cause paralysis of muscles by blocking the nervous impulses to them. It very often causes death. Fortunately it is a rare disease.

Incubation Period

The toxin is already present in the food when swallowed and it rapidly passes through the wall of the stomach and into the blood-stream that carries it to nerve-endings in muscles. The incubation period is commonly between 10 and 12 hours: it may be shorter or longer, probably depending on the account of toxin in the food.

Symptoms

The first symptoms are those of vague discomfort and weakness. Soon signs of paralysis occur. The first is blurring of vision due to weakness of the muscles that help the eyes to focus. Next drooping of the eyelids occurs and difficulty in speaking is felt. The muscles of the throat are soon affected and the patient becomes unable to swallow even his own saliva. The breathing muscles become paralysed, and the condition then is one of great distress. The patient's brain remains clear and he is fully aware of his desperate plight. After this the course is rapidly downhill. Very few, if any, who reach this stage, recover, but some patients, who have eaten very little of the food, escape with mild attacks.

Diagnosis

If several patients fall ill at the same time with these symptoms, the diagnosis of botulism is obvious, for there is no other illness capable of causing simultaneous and rapid paralysis in a group of people. If there is only one patient, it is difficult to diagnose, for many other illnesses can produce solitary cases of paralysis. Specimens of vomit, faeces and of food are examined in the laboratory. The germs may grow on special media, and the presence of toxin may be shown by injecting extracts of the food into mice.

7.6 *Dysentery*

Dysentery is not typically food borne disease but occasionally it can be. The symptoms may be indistinguishable from those of salmonella food poisoning. There is diarrhoea, some abdominal pain and sometimes vomiting. Shigellae sometimes cause ulceration of the bowel wall, and when this occurs there may be bleeding from the bowel. Usually this is very slight for most causes of dysentery are mild; sometimes the attack is severe and the patient suffers greatly from loss of blood. A good deal depends on which germ causes the infection; *Shigella sonnei* causes mild illnesses, whereas *Shigella shigae* tends to cause severe attacks.

There are many other illnesses that can be conveyed by food. Typhoid fever, brucellosis (Malta fever), and infections hepatitis (jaundice) are examples, but these are not included under the term food poisoning. The food handler should know of their existence. The methods of preventing food poisoning will also prevent these other diseases from being conveyed by food.

7.7 *Other Forms of Food Poisoning*

Other forms of food poisoning are caused by chemicals or impurities getting into sound food, or when we eat, by mistake, substances which are poisonous but resemble edible foods, the commonest of the former type is metallic poisoning, and of the latter type the best known is mushroom poisoning.

7.8 *Metallic or Chemical Poisoning*

This may occur when from a pan or container seeps into the food. Usually the food is acidic and the pan is made either of galvanized iron or cheap enamel that has become chipped. Outbreaks of zinc poisoning have taken place when stewed apples or other acid fruits have been prepared in galvanized iron pans. The commonest symptom is severe vomiting and this may occur within minutes, or after an hour or two, depending on the concentration of zinc in the food. In one outbreak, alcoholic punch was stored in galvanized containers for over 50 hours and contained zinc in a concentration of 2,200 parts per million: many of those who drank it were vomiting within 20 minutes and nearly all the guests were ill within the hour. Sometimes severe diarrhoea is the most prominent symptom, as in one outbreak caused by a meal of chicken and spinach which had been stored in a galvanized tub after preparation. In another outbreak, iced lollies caused zinc poisoning, and in the past, antimony poisoning was caused by storing lemonade and other acid drinks in chipped enamel jugs.

Metallic poisoning has not so far been reported from the use of aluminium, iron, tin or stainless steel pans or containers, and zinc poisoning from galvanized pans is now quite rare, for the danger is well–known. When metallic or other chemical poisoning takes place it usually follows some accidental contamination. In one outbreak potassium bromate got into sugar used in cakes, and, in another, barium carbonate was used instead of potato starch in a sausage mix. Copper poisoning has been caused by making coffee with water from a dilapidated electric boiler. However, all these forms of chemical or metallic poisoning are rare indeed, and provided food is prepared and stored in sound equipment the danger of such poisoning can be almost disregarded.

Unusual and sometimes very serious forms of poisoning have resulted when food being carried in a van or a ship has been accidently contaminated with a chemical that has leaked from another part of the cargo. Four serious outbreaks of food poisoning were caused in this way in Saudi Arabia in 1966. Flour had been shipped from America to Arabia in two ships, and in the same holds, above the flour, was a cargo of liquid endrin, a poisonous insecticide. Some of the liquid leaked and soiled the bags of flour, and when this was made into bread, 874 patients were admitted to hospital with a severe from of food poisoning and 26 patients died in coma. International regulations have since been drafted to prevent such errors in the loading of ships, but tragic mistakes may still be made us unforeseen dangers are revealed. In one in Epping in 1966, 84 patients suffered from jaundice after eating wholemeal bread. The flour had been carried in a van in which there were also jars of a chemical used as a hardener of resin. The liquid had spilled and though there were only traces of it in the flour, this was enough to cause a serious epidemic.

7.9 Mushroom Poisoning

This name is unfortunate, as there is no such thing as poisoning due to mushrooms. Mushrooms are delicious and wholesome food, as also most fungi found in our woods and meadows, even those we call toadstools; but there are several fungi that contain deadly poisons. Such are the 'fly agaric', *Amanita muscaria*, the 'false blusher', *Amanita pantherina*, the 'death cap', *Amanita phalloides* and the 'destroying angle', *Amanita virosa*, Muscarine and amanitine are two of the poisons concerned and they cause severe vomiting and diarrhoea and often coma and death. Such fungus poisoning is, however, uncommon, for these deadly fungi can be distinguished easily from the edible mushroom. The fly agaric, for example, is bright scarlet with white spots; the 'death cap' is greenish with white gills; the 'false blusher' brown with white spots; the 'destroying angel' is white all over, even its gills; whereas the gills of the edible mushrooms are either pink or brownish-black. The skin of the cap can be peeled easily from the edible mushroom, but this peeling test is a very unsafe one, for some of the poisonous fungi peel easily too. The only safe way, if one wants to be adventurous in eating fungi, is to learn most carefully the distinctive features of the really dangerous ones—there are less than a dozen. Most other fungi can be eaten: a few may cause slight gastro-intestinal upset, without being dangerous, but most are safe and delicious.

7.10 Mussel Poisoning

While all these forms of poisoning are caused, to some extent by man's carelessness or ignorance, nature itself can sometimes lay poisonous baits for man. Mussels are normally a delightful food, to those who like them. They feed by straining huge volumes of sea-water through their bodies and sifting out the microscopic specks of animal and plant life in it. Sometimes a population explosion occurs in this microscopic world, and vast multitudes of organisms known as dinoflagellates spread over the surface of the sea forming a red fluorescent film. These dinoflagellates produce a poison that does not harm the mussels, but can cause a severe illness in man if he eats them. There was a serious outbreak of this form of poisoning in North-East England in the summer of 1968. At least 78 people were severely affected with peculiar nervous symptoms, including paralysis; but fortunately none died. There was, however, a great number of deaths among shags and other sea birds that fed on sand eels that were also poisoned by the dinoflagellates, and many starfish that fed on mussels were found dead on the shore.

There is little the food hygienist can do to avoid this form of poisoning. For this one must look to the biologists and naturalists who could test mussels as the summer advances for the presence of toxin; and dead birds on the sea shore might also give a warning; but these dinoflagellate explosions are rare occurences, freaks of nature, and one can normally eat one's mussels without fearing any consequences.

7.11 Ptomaine Poisoning

This term is mentioned because the name still creeps into accounts of food poisoning, especially in the lay press. There is really no such thing. Ptomaines are substances formed in putrefying organic matter such as putrid food. But ptomaines are not poisonous and, in any case, no one eats food so putrid. Food which results in food poisoning is usually fresh and attractive. It is the invisible germs, toxins or chemicals that cause the poisoning. The food itself has not 'gone off' or become tainted. 'Sweet tastes have sour closes.'

8

Some Other Diseases Spread by Food

8.1 Introduction

Any germ that can survive in food can be conveyed to man by food. It is merely convention that labels only certain food-borne illnesses as food poisoning. Some of the other diseases conveyed by food are much more serious than food poisoning.

8.2 Typhoid Fever

This disease is caused by a salmonella known as *Salmonella typhi* but the disease is very different from salmonella food poisoning. The typhoid salmonella does not stay in the bowel like most other salmonellae, but enters the blood-stream from where it can penetrate almost any organ of the body. The patient suffers a severe prostrating illness with prolonged fever, and there are various dangerous complications. Fortunately an antibiotic known as chloramphenicol has a dramatic effect on this disease. If given early, it can bring the illness to an end in a few days and the patient regains his sense of well-being in a week or ten days, instead of suffering a long, muttering, feverish illness for weeks together.

Typhoid fever is always caught through food or water. The dose of germs may be quite small. Other salmonellae must multiply to great numbers in food before they cause food poisoning, but a few typhoid salmonellae can cause typhoid fever. This is why typhoid fever can be conveyed by water. Germs do no readily multiply in cold water though they can survive in it for weeks together. Big outbreaks of typhoid fever have been caused through drinking water, but there has never been a water-borne outbreak of salmonella food poisoning.

Salmonella typhi differs from most other salmonella in that it is a parasite of man only. Other salmonellae may be found in creatures as diverse as foxes, pythons, turtles, pigeons and many others as also man. Not so *Salmonella typhi*; it infects only man. This means, that every case and every outbreak of typhoid fever must originate from typhoid germs passed from another human being. This means a typhoid carrier — a man or woman who has recovered from typhoid fever but still passes the germ in his or her faeces or urine. Some carriers continue to do so for many years, even for life. Somehow the germs must find their way into water or food, and this water or food must be swallowed by someone else.

Though there are alarming reports about cases of typhoid fever in the press and on radio and television, but typhoid fever does not pass through the air from person to person like measles or influenza. Unless the germs are present in the food or water supply, typhoid fever will not spread, and unless a person swallows water or food with typhoid germs in it, he will not get typhoid fever. The germs get into food generally directly because the carrier is a cook or a food handler. The way they get into water supplies may be much more involved and the unravelling of an outbreak can involve an intricate piece of detective work.

8.3 *Infectious Hepatitis*

Infectious hepatitis is jaundice. *Hepar* is the Greek for liver and hepatitis means inflammation of the liver. When the liver is inflamed, bile cannot flow through it so that it gets dammed back into the blood-stream: from there is gets caught in the cells of the skin and conjunctiva, and turns the patient yellow.

The disease is caused by a virus but no one has yet been able to grow this virus. This makes it difficult to be sure of how the disease spreads. It seems to spread from person to person in families and class-rooms, and it can also spread from house to house, as if carried by travellers over hill and dale. Usually one case is separated from the next by a month, so that the disease seems to have a long incubation period. Sometimes, however, explosive outbreaks take place and a lot of people go down with jaundice within a short time. This can only mean a common source of infection. In Delhi in 1955 over 29,000 people were affected within 6 weeks and there is no doubt that the outbreak was due to massive contamination of the drinking-water supply. In another very small outbreak, doctors and nurses who had orange juice for breakfast developed jaundice: the orange juice was prepared by a kitchen worker whose husband also had jaundice. In another outbreak in an officer's mess a potato salad seemed to have carried the virus: it had been prepared by a kitchen hand who had himself recently been ill. The eating of shellfish, especially clams, has often been connected with outbreaks in America. In Scandinavia an outbreak of infectious hepatitis was traced to the eating of oysters, and the supplier of the oysters was sued for damages. It seems that food and water play an important part in the spread of this form of jaundice, but just how important it is difficult to judge as we have still no way of getting the virus to grow in the laboratory.

8.4 Scarlet Fever and Diphtheria

Scarlet fever is not a very common disease in Britain today, though in the past there have been big and serious outbreaks. Sometimes these have been conveyed by milk. This may occur in one of two ways. A milker or other dairy worker may be suffering from mild scarlet fever or tonsillitis or be a carrier of the germ known as haemolytic streptococcus that causes both these conditions, and germs from his throat, nose or saliva may get into the milk. This is not a very likely method of spread for these germs do not readily multiply in milk at room temperature. A more likely source of a milk-borne outbreak of scarlet fever is inflammation of the udder, or mastitis, in the cow. This is a fairly common condition in cows: it is not usually caused by the scarlet fever streptococcus, but it can be. The germs multiply in the inflamed udder and when the milk is passed it is teeming with them. This has been the usual story in milk-borne outbreaks, but pasteurization kills streptococci, and outbreaks of scarlet fever cannot be caused by pasteurized milk.

Diphtheria is a disease spread from one person to another, for the germ is present in the throat and nose of a patient with diphtheria or of a diphtheria carrier. There have in the past been one or two outbreaks conveyed by milk. The milker has generally been a carrier, the diphtheria germ has been conveyed on the hands of the milker to the teats of a cow, a diphtheritic sore has developed on the teats and the germs have passed into the milk. The milkman has then carried the germ in the unpasteurized milk to the customers on his rounds. Pasteurization has minimised this risk, but it was never a serious one: a milk-borne outbreak of diphtheria was an epidemiological curiosity even before the days of pasteurization.

8.5 Brucellosis

The germ of this disease is known as brucella, named after Dr. Bruce who first discovered it when investigating outbreaks of the disease in Malta. The two commonest strains are *Brucella abortus* and *Brucella melitensis*. The former causes a disease in cattle that often ends in abortion. It is quite common in Britain. The latter causes a wasting disease in goats and sheep, mainly in Mediterranean countries. The infection is conveyed to man either by drinking the milk of cows, goats or ewes or by direct contact with the animals, especially during calving, kidding and lambing. It is a common infection in veterinarians in Britain and also in goat-herdsmen and shepherds in Mediterranean areas.

Symptoms

The disease consists of a long feverish illness with drenching sweats and pains in the bones and joints. Sometimes the fever comes in bouts or waves and the illness is then called undulant fever. *Brucella melitensis* causes a more severe illness than *Brucella abortus*. It was till recently very common in Malta and was known as Malta fever. The peasants there used to lead their goats into the towns of Floriana and Valetta and milk them at the customers' doors, a very direct producer to consumer service but one that was full of danger to both. Pasteurization is now compulsory in Malta and the disease has almost disappeared, though cases are still found among the peasants who live close to their goats and probably still drink milk raw. In Britain the disease is caused by *Brucella abortus* and infection is derived from cows only. Cases are still found in areas where unpasteurized milk is drunk but it is commonest in veterinarians and farmers who come in close contact with the infected cattle and with the milk before it is pasteurized.

Pasteurization, of course, can only kill the germ after it has passed through and caused illness in the animals. The only final solution to the problem of brucellosis is to carry out tests for brucellosis on all the animals in herds, slaughter those that are positive, and so raise disease-free herds. This would be a formidable task but it is far from impossible with cattle in Britain. It has been done in many European countries and there is now a campaign to deal with the infection in Britain too. After all, this was the policy carried out when the problem of tuberculosis in cattle was tackled. The orthopaedic hospitals used to be full of children suffering from tuberculosis of the bones, joints and spine, all caused from drinking the germs of tuberculosis in milk. We now have tuberculosis-free herds. It is time to get rid of brucellosis too.

Diseases Spread by the Water and the Air

8.6 Water

All food contains water. The preparation, manipulation and cooking of food involves its use. We even drink water. What *is* this substance? To answer this question we start with the *sources of water.*

The source of our water-supply is rain, which falls from the clouds and soaks into the soil. Were there nothing to stop it, it would sink deeper and deeper and form a vast subterranean ocean of water that we tap only by enormous bore-holes and wells, but the water is stopped by the rocks under the soil. Sometimes the first layer of rock beneath the top soil is watertight or impermeable, and the rain can sink no further. It therefore collects on top of the rock as sub-soil water, and as the layer of rock is seldom horizontal, this sub-soil water flows under the surface of the earth according to the tilt of the rock. Eventually it forms a basin of water which somewhere, generally at the foot of a hill or mountain, breaks surface and gushes out as a spring. We think of a spring as something tiny and secluded, but all streams and rivers have their sources in springs, which must then be enormous. A *spring* is the natural way by which rain comes to the surface, but man can also reach the sub-soil water by sinking a well and drawing the water up. Such a well may be very deep or quite shallow. It depends how far below the surface is the impermeable first layer of rock, but whether it is deep or shallow, a well which taps the sub-soil water is always known as *a shallow well.*

Sometimes the first layer of rock is permeable; the water then sinks through the rock, perhaps through two or three layers of rock, till it is stopped by a deeper impermeable layer. Man can sink a well by boring down through these layers to tap this source of water. Such a well is known as a *deep well* because it is sunk through rock. It may or may not be really deep. That depends on the depth of the first impermeable layer, and a deep well in one part of the country might be shallower than a *shallow well* in another. Note the difference between the impermeable first layer and the first impermeable layer.

There are two more important differences than depth. Water is a powerful solvent, and as it sinks through rock, it may dissolve out some of its substance. If the rock is limestone, the water dissolves some of its calcium, and calcium makes water hard. Deep-well water is therefore often hard, while shallow-well water is soft, as it has not passed through rock. The second point is that a rock, even if it lets water seep through, also acts as a filter, and many impurities including germs, are filtered out of the water as it sinks through the rock. Deep-well water may therefore be purer than shallow-well water, but the latter is softer.

Rivers, Streams and Lakes

Rivers and streams, are fed by enormous springs. They are really huge masses of water that have escaped from the sub-soil and are moving down an incline to the sea. Lakes are also collections of sub-soil water: some are fed by steams and rivers, while others form on high ground with no visible source of supply: they are simply huge springs. The water in the rivers, streams and lakes, when used as a water supply, is known as upland-surface water.

How Safe is Water?

As the rain falls through the air it may dissolve traces of impurities, like sulphur, from our smoky atmosphere, but it is free from germs and safe to drink. Once it starts sinking through the soil, its safety can no longer be guaranteed, since the soil is full of germs that can be washed into the sub-soil water. Most of these germs are harmless to man, but disease can get into the sub-soil water also. This can happen, for example, if there is any defect in a drain or sewer that allows sewage to escape into the ground, for this is bound to flow into the sub-soil water and carry any germs with it. Septic tanks and cesspools often leak and may cause serious contamination of the sub-soil water with dangerous germs. On farms there are always heaps of animal manure, and liquid drains from this into sub-soil water. Thus it is important that shallow wells should be so sited that impure water from such sources cannot drain into them. Yet many outbreaks of disease have been caused in this way.

Rivers and streams may be polluted by everything that flows into them. This includes waste products from factories, but also water from countless rivulets, ditches, drains and effluents. Every agricultural field has drains under it. He spreads fertilizers, pesticides and manure on his fields, and seepage can and does get into the rivers. More important is the drainage or effluent from sewage works and septic tanks, as these generally contain disease germs, and these germs get into our water supplies and must be destroyed before we use it. If they are not destroyed disaster follows if the water is used for cooling cans of meat in a canning factory or for drinking. This talk on wells and water supplies may seem a long way from food hygiene, but there is scarcely any part of food preparation that does not involve the use of water. It cannot harm a chef, a shopkeeper or a housewife to know where the tap-water comes from and what it is. Let us look quickly at sewage works and waterworks so that we can understand more clearly just what comes out when we turn on the tap.

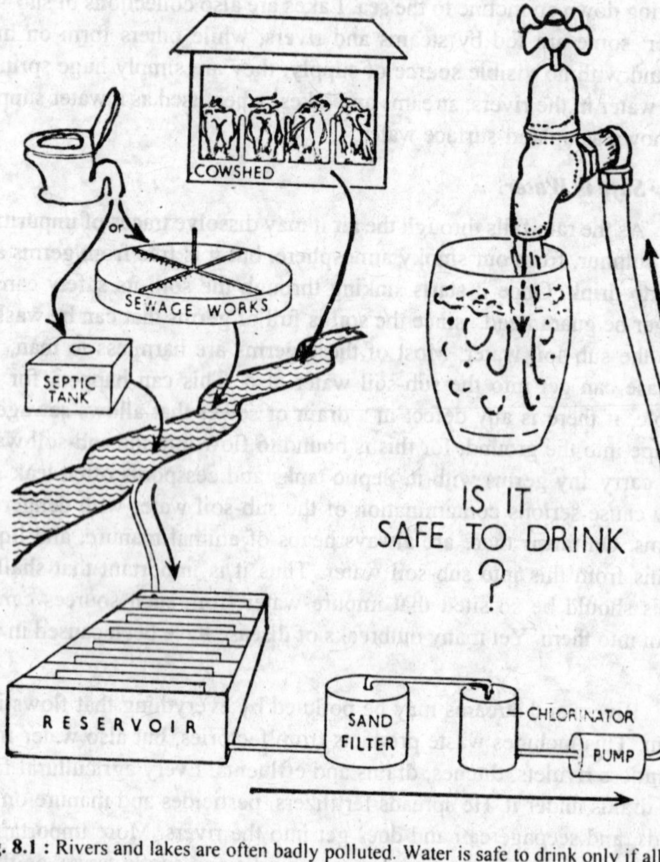

Fig. 8.1 : Rivers and lakes are often badly polluted. Water is safe to drink only if all precautions are taken to remove dangerous germs from it before it reaches our taps.

8.7 Sewage Works

Most of us think the purpose of sewage works is to destroy germs. However, the actually this is not so. Their purpose is to cut down the amount of organic matter flowing into our rivers. This is achieved first by removing most of the solid matter by screens or coarse filters, and secondly, by exposing the liquid part to both aerobic and anaerobic bacterial action. This is achieved by allowing the sewage to trickle out of rotating sprinklers and sink down through filter beds. The action of the air and the bacteria is to break down the organic particles into simpler substances. This process needs a lot of oxygen, and the aim of the sewage treatment is to decrease this biochemical oxygen demand (the B.O.D.) of the crude sewage before it goes into the rivers. If the sewage were discharged untreated into rivers, it would use up all the oxygen dissolved in the water, and water plants and fish, bereft of oxygen, would die. Our rivers would be aesthetically more unpleasing than they are now, but the primary purpose of sewage works is to save river oxygen, not to please the eye.

During sewage purification some disease germs may die, but many do not. If there are cases of typhoid fever or poliomyelitis in a town, the germs of these diseases will almost certainly be found in the effluent flowing from the town's sewage works where it runs into the river. Carriers of disease germs have been detected by soaking swabs in the effluent and then trying to trace back from there by drain swabs to the water closet of the carrier.

8.8 *Waterworks*

Waterworks take their water straight from rivers and lakes. Now disease germs do not find water a very nourishing fluid, and so one of the main purifying methods in a waterworks is simply to hold the water for several weeks in an attempt to starve the germs. A reservoir is not so simple as it looks on the surface. There are long walls or baffles in it and water coming in at one end has a long, tortuous journey before it gets out at the other. It probably takes about three weeks, and in this time a lot of fine suspended particles will settle to the bottom and many germs will die: but not all. Typhoid germs and many viruses can survive these bleak weeks. The water coming from the reservoirs is therefore led through very fine sand filters which remove any particulate matter still in it and also many germs, and, after filtration, it is usually treated with chlorine which kills most of those that survive to this stage. The treated water then goes to the pumping stations and is pumped into our water mains and so into our taps. It is a long process, and obviously a fault anywhere along the line could lead to trouble. Our water engineers are constantly watching and improving the processes: our bacteriologists and our virologists are constantly testing the final product.

This may seem a long diatribe about something rather remote from food poisoning. The thoughtful food hygienist will realise how much he depends on a safe water supply.

8.9 Air

If a bacteriologist leaves a plate of culture medium exposed to the air, germs will grow on it. They can only come from the air. A hospital bacteriologist often does this very thing when he is trying to trace the spread of infection in a ward or operating theatre. A public health bacteriologist may do exactly the same when investigating outbreaks of food poisoning traced to a bakery, kitchen or a factory. Air is not germfree, certainly not the air inside a building. It carries germs breathed out by people when they speak, and especially when they cough or sneeze. It carries germs stirred up from the floor during sweeping, from people's clothes and boots as they walk, and from any powder or dusty material, such as flour or dried egg, that may be used in baking processes. Moulds and yeasts float through the air. If they land on bread or cheese, they will soon cover it with greenery: if they alight on fruit or juices, they start fermenting them, changing the sugar in them into alcohol. If there are disease germs anywhere, on the hands, boots, clothes or hair of a person, or on any food, or other material he is working with, these germs can easily be wafted into the air and spread all over the place. Thus a food hygienist should see that everything is done to prevent germs getting into the air, and to keep food covered and protected so that if there are any air borne germs around, at least they cannot come down on the food.

9

The Trail of Infection

9.1 Infection on the Farm

Farming is fundamentally an unnatural process. It consists of controlling the growth of plants and animals, encouraging those features that promises a good return in the market, and pruning out those that are unprofitable. Animals that in nature would roam widely in search of food and mates are, on farms, confined to fields or locked in buildings, their reproduction regulated by artificial insemination and their milk production controlled by dairy requirements. A cowshed is not the natural environment for the cow, nor a deep-litter shed for a hen. Yet under these unnatural conditions the animals thrive, for they get more food, water and shelter than they could obtain in the wild state, but if some foreign disease-germ breaks in, the artificial surroundings of their domesticated reserve at once become highly dangerous.

Animals defaecate and urinate where they stand, and no matter how well washed and clean a cowshed may be, the floors, the bedding, the stalls and the walls can never be sterile. There must always be a warm, moist film of organic material on all surfaces, and pathogenic germs can flourish in these conditions. The very processes of cleaning with brushes and water may stir this material up and cause an invisible mist of infectious material in the atmosphere. Soon the cowshed can be a very dangerous place for the cows. Germs may settle on their foodstuff or be present in the air they breathe. Shortly some of the animals become sick, and sick animal in turn passes the germ in its faeces, urine, breath or milk and so adds to the weight of infection in the cowshed.

9.2 Source of Infection

In animal like in human beings, outbreaks of disease occurs and it is always difficult to determine how the infecting germ gets into a community: once it is in, it is much easier to trace its spread. There are two main avenues of infection on farms; one is the introduction of new animals, and the other is artificial foodstuffs.

Animal Carriers

Cattle : When a cow gets infected with a salmonella, it may suffer from severe diarrhoea or have a vague feverish illness. This is generally called 'cattle paratyphoid', but it is not caused by *Salmonella paratyphi*, that causes paratyphoid in man; it is caused by *Saslmonella dublin*. Some cows die of 'cattle paratyphoid' but most recover: though they become perfectly fit again, they continue to pass the germs in their faeces, mostly for the rest of their lives. Thus animal becomes a carrier: a carrier excretes the germ, but suffers no ill effects from it after the original illness. It is, however, a great danger to other animals. If such a cow is bought at market and added to a 'clean' herd, it will sooner or later cause infection in the other animals.

Calves also suffer from salmonella infection, generally in short, sharp epidemics. Under intensive methods of stock farming, calves from hundreds of different farms may be brought together in one unit for rearing. The calves are separated from their dams a few hours to a few days after birth. They are taken in trucks to markets or dealers' premises and the trucks pick up calves from other farms on the way. They are then sold to rearing farms where they are kept for different periods. Some rearers move them at three months to other farms for fattening, others keep them till they are slaughtered at 12 to 15 months old. An intensive calf-farming unit of 500 calves may get its stock from over 400 different farms so that a great deal of mixing occurs. Thus, if a few calves arrive in an infectious state the infection can spread rapidly. The illness is generally milder than in adult cows and the calves do not becomes permanent carriers, but excrete the germ for only a few weeks, but sometimes the infection takes a more severe form and many calves die. If the germ is *Salmonella dublin* the origin of the infection is probably an adult cow on one of the farms, for *Salmonella dublin* is a parasite mainly of cattle If another salmonella is the cause, say *Salmonella typhimurium*, then the infection may have come from other sources—rats, fowls or feeding stuffs—for *Salmonella typhimurium* infects many different hosts.

Pigs : All the above relates to cattle, but other animals on the farm suffer similar infections. Pigs are generally infected with salmonellae: the pig is the main host of *Salmonella cholerae-suis*, but it is also infected with other salmonellae as well, such as *Salmonella typhimurium* and *Salmonella dublin*. Pigs very readily infect one another and the germ, after the acute attack, settles down in the abdominal glands of the pig and is often still present there when the animal is slaughtered for food.

Sheep: Sheep are rarely infected with salmonellae. They live out in the open, instead of in cowsheds or sties, so that the germs have less chance of building up a heavy load of infection in the environment. But salmonellae can live in soil and pasture: they have been grown from cow-dung that has lain for 9 months in a field.

Poultry : Farmyard fowls are generally heavily infected. They have two salmonellae of their own, *Salmonella pullorum* and *Salmonella gallinarum*, and these have very seldom caused infection in man. Fowls may, however, be infected with many other salmonellae. Often they show no sign of illness but can still spread the infection to other animals on the farm. Hens, ducks, geese and turkeys may all be infected. The salmonellae are often still present in the carcass of fowls when killed for food, and if the bird is not properly cooked, those who eat it may get food poisoning.

Rats and Mice : Rats and mice become heavily infected if there are salmonellae on the farm and they can spread the infection by contaminating feeding stuffs, bedding and drinking water. However, they are really the victims, not the cause of the infection, and if the farm is cleaned up and infected stock animals removed, the infection dies out in the rats and mice, but it is better to get rid of the rats and mice too.

Dogs and Cats : Dogs and cats wander round farm buildings and pick up and eat all kinds of infected material. They may become infected with salmonellae and then carry the infection into the farmhouse and infect the family. A member of the family may become a carrier, and he or she can spread the infection, and perhaps contaminate milk or other farm products. It is not a common occurrence, but outbreaks have been caused in this way.

Feeding Stuffs : Feeding stuffs are supplied generally as meal or pellets, and always contain animal protein. This protein comes from the carcasses of dead animals and these animals may have been infected with salmonellae before death. If the germs survive in the feeding stuffs and are conveyed to other animals, perhaps across an ocean, for much animal feeding stuff is imported. As much as 15 to 45 per cent of artificial feeding stuffs has been found to be infected with salmonellae. Man is not the only animal that suffers from food infections: the animals which are fed on these feeding stuffs are subjected to a compulsory form of food poisoning. It is better to sterilize feeding stuffs before they go on the market.

All the above has dealt with salmonella infection on the farm. Staphylococcal infection of milk is another hazard on the farm, although a minor one.

9.3 Animals on the Move

Crowd Diseases

Epidemiology deals with the study of crowd diseases. The word comes from the Greek, *epi* which means upon, and *demos* meaning the people or the crowd, so that an epidemic is something that falls upon the crowd. The reason is, of course, that when men travel rough, living standards fall: feeding facilities are improvised and sanitary hygiene is primitive. Moreover, the men are exposed to great stresses and strains: they become over-tired and dispirited and, if infection breaks out, they are less able to resist it in their weakened state.

Stress and Shrinkage

Animals on the move suffer too. When they leave the farm they may be driven on the hoof for long distances, or be huddled together in lorries or train trucks for hours on end. Perhaps they are not fed and watered properly, and they must suffer from excitement and fear. When they reach the market, they are driven into pens and later into sale-rings, where they are paraded and prodded before auctioneer and farmers. All this produces *stress*, a word used by vets to describe the strains which cause animals to lose condition while on the move, Farmers refer to *shrinkage*. This means the loss of weight that occurs in their animals between the farm and the sale-ring. Even under the best conditions of transport, the animals obviously suffer. How do they fare from infection?

From Farm to Market

When a truck leaves the first farm it may have only a few healthy animals on board. It stops at other farms on the way, loading a few more at each stop. Most of these also are healthy, but one or two may be infected. By the end of the journey the truck is soiled with faeces, the animals have been badly jostled, and any germs that are present in one animal have had every chance to spread to the others.

At the market other trucks are arriving, bringing cattle, sheep and pigs from farms scattered far and wide. The animals are unloaded and herded into pens and lairs in the market buildings where they may spend hours or days before being moved on to new farms, or to the abattoir. The bedding, the floors, walls and partitions are spattered with dung and urine, the temperature is warm and the atmosphere moist. The conditions are ideal for the multiplication and spread of germs.

Once walls and floors and partitions are contaminated it is not easy to make them free of germs. If these structures become caked with animal faeces, salmonellae and other germs can live almost indefinitely. To avoid this, absolute cleanliness with adequate and frequent hosing is needed, but often not provided, and pens and lairs, as distinct from farms, may become hot-beds of infection. Animals coming into them have little chance of avoiding infection. Many investigations have been carried out by bacteriologists to show that infection spreads in this way, and that the longer the animals are held in such pens and lairs, the greater is the rise in infection.

At the Abattoir

Animals arrive at the abattoir straight from farms, from markets where they have been sold for slaughter, or in ships from overseas. They are led into pens in buildings known as lairages, there to await their final journey to the adjoining abattoir floor. In the pens they are exposed to the same danger of infection as in the markets, and if the lairages are badly managed, animals can become infected in their last few days of life. Even after slaughter the danger is not because clean carcasses can be contaminated with germs from the carcasses of infected animals. The slaughter-house side of an abattoir is generally heavily contaminated with germs, and salmonellae have been grown from floors, dehairing machines, knives, chain tackle, trolleys, trucks and table-tops. It is easy to see how germs can spread under such circumstances, and a side of beef from an animal that was clean and healthy on the farm may leave the abattoir covered with pathogenic germs it has collected on the way.

Poultry Plants

Hens, ducks, geese and turkeys go to poultry plants and not to abattoirs. They undergo similar stresses and probably more chances of infection on the way, but they are not held, like the animals, in lairs before slaughter. At the poultry plant they are quickly killed, bled, defeathered, eviscerated, trussed and packed, all on a conveyor belt system. Along the line, contamination may build up anywhere and germs can be grown from evisceration and giblet troughs, chilling tanks and rinsing water: they can generally be grown from the floor and from drain swabs. Most important of all, disease germs may be present in the trussed fowl, either because the bird was already infected before slaughter or because the carcass has been contaminated on the way through the plant.

It is a long story from the entry of the germ into a farm to the exit of contaminated food from an abattoir or poultry plant, but germs have a tenacious hold on life and can adapt themselves to many changes of environment. Under favourable conditions they can survive on dead meat for long periods and, if a little warmth and moisture are available they can soon multiply and be ready to take up a new parasitic existence inside the body of the human being who swallows them in his food.

It must not be thought that every market, every abattoir and every poultry plant is filthy and full of disease germs. Moreover, cleanliness is not the same thing as asepsis or freedom from germs. Disease germs spread even in quite clean farms.

9.4 From Abattoir to Factory

Most of us are familiar with the products—the side of beef, the joint, the chicken or the turkey. All this is normally very wholesome food, but, as we have seen, this kind of food is attractive not only to man, but also to microbes. What happens when carcasses of animals or birds leave the abattoir or poultry plant contaminated with disease germs?

Meat Factories

Sausages : The journey of germ that enters a meat factory on a side of beef depends on the processes carried out there. If sausages are produced, the meat is chopped up and put in a mixing machine to which meal and spices are added. The sausage meat coming out of the mixer is then fed into a filling machine and sausages emerge from the other side. These are then carried on conveyor belts to a packing machine where the sausage are enclosed in Cellophane packets, untouched by human hands. The whole process can be very clean, but the danger of infection is unseen. If the surface of any carcass meat from the abattoir is contaminated with salmonellae, these germs are thoroughly spread around in the mixing process. All the plant becomes contaminated, and sausages made from wholesome meat are liable to contain salmonellae by the time they reach the packing machine. Moreover, the salmonellae are now *inside* the sausages, not just on the surface. This is important, for in lightly fried sausages the centre may easily not reach a high enough temperature to kill salmonellae. Salmonellae have, in fact, survived in the centre of sausages gently fried for 15 minutes. It has been calculated that about 4 per cent of sausages sold in Britain contain salmonellae. This is a low percentage, but as around 9,000 million are sold every year, it represents a lot of sausages. They should all the browned on both sides.

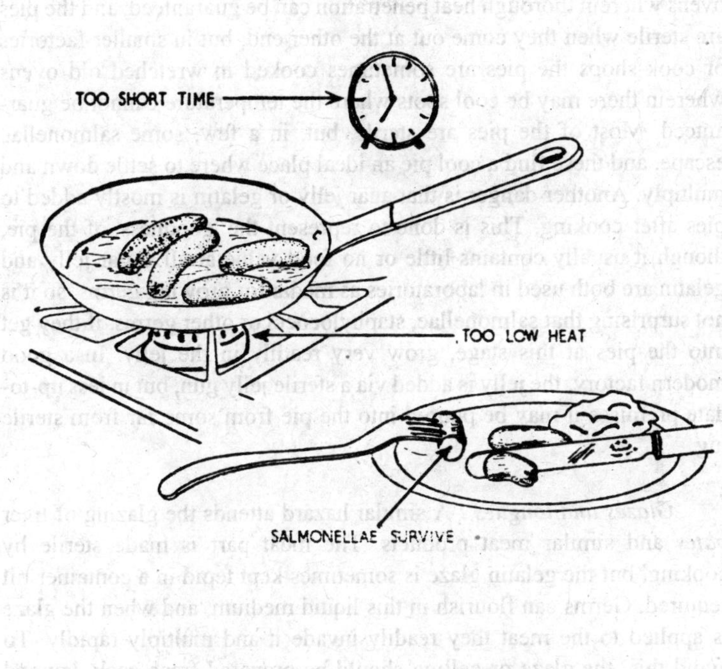

Fig. 9.1: Salmonellae can survive gentle frying *inside* sausages; brown them well on both sides

Cooked meat : Often joints of meat or pork are roasted in meat factories and sent out cooked to wholesalers and retailers. Such meat may be contaminated when it arrives from the abattoir, but the germs, being surface germs, are sure to be killed in the roasting. The danger is that tables and other surfaces in the factory may be contaminated with salmonellae from the raw meat. If meat is placed on these after cooking, it can pick up some of these germs and, if the temperature is right, they can multiply on the surface of the cooked meat and eventually be swallowed by unfortunate consumers. This is not a theoretical risk, but one of the fundamental dangers of bulk food production. *Raw and cooked food should never be handled on the same surfaces*.

Pies : Pies are often made in the same factory as sausages and the same risks are involved, for any germs on the *surface* of abattoir meat are tucked well inside the *centre* of a pie. In a good modern food factory this should not matter, for the pies are cooked in magnificent automatic ovens wherein thorough heat penetration can be guaranteed, and the pies are sterile when they come out at the other end; but in smaller factories or cook-shops the pies are sometimes cooked in wretched old ovens wherein there may be cool spots where the temperature cannot be guaranteed. Most of the pies are sterile, but, in a few, some salmonellae escape, and these find a cool pie an ideal place where to settle down and multiply. Another danger is that agar jelly or gelatin is mostly added to pies after cooking. This is done to represent the 'goodness' of the pie, though it usually contains little or no food value at all. Agar jelly and gelatin are both used in laboratories as media for growing germs, so it is not surprising that salmonellae, staphylococci or other germs, if they get into the pies at this stage, grow very readily in the jelly. In a good modern factory, the jelly is added via a sterile jelly gun, but in less up-to-date premises it may be poured into the pie from some far from sterile jug.

Glazes and tongues : A similar hazard attends the glazing of liver *pates* and similar meat products. The meat part is made sterile by cooking, but the gelatin glaze is sometimes kept tepid in a container till required. Germs can flourish in this liquid medium, and when the glaze is applied to the meat they readily invade it and multiply rapidly. To avoid this, the glaze or gelling should be prepared fresh each day and boiled for at least 5 minutes. It should be kept, while being used, at a temperature of at least 72° C. (161.6° F.). Any not used should be cooled rapidly and kept all night in a refrigerator, and boiled again next day, but it is much better to throw away any surplus glaze.

Tongues must be partially cooked before they can be skinned. This involves handling, and germs can easily be added at this stage. The skinning should be done with the tongues as hot as can be borne, and the cooking completed immediately after, so that any added germs are killed before they have time to multiply. The tongues can then be cooked and pressed in sterile containers.

Poultry barbecues : Hens, ducks, geese and turkeys are generally heavily infected with salmonellae, but food poisoning outbreaks from eating poultry have been unusual in Britain. This is because birds have generally been bought fresh and cooked in the oven in the home. Fashions change in cooking as in other things, and this has been particularly the case with poultry, especially chicken. The carcasses, instead of being sold fresh, are now often deep-frozen, and sent in the frozen state direct to retailer or restaurants. Here we must recall that *freezing does not kill germs* — it merely stops them multiplying. If salmonellae are present in a bird when it goes into a deep-freeze, they are still there and alive when it comes out. Now it takes up to 12 hours to thoroughly thaw a big bird. If the time is greatly cut down, the centre of the bird may still be frozen when cooking begins, and due to this the temperature in the centre may never be high enough to kill salmonellae, even though the outside of the bird is beautifully browned. This is liable to occur in spit rotisseries and chicken barbecues, for three reasons: (1) that there is often limited space, and therefore it is difficult to lay out a lot of frozen birds to thaw, (2) that the owner may be in a hurry to get the birds ready for sale, and (3) that in spit roasting, with repeated opening and closing of the doors, the temperature is never high enough to penetrate the deepest part of the carcass. Outbreaks of food poisoning have been caused in this manner and they may become commoner.

A second danger with cooked poultry is that it may become contaminated after cooking, in the same way as cooked-meat products, by chopping and preparing the cooked bird on the same blocks as the raw birds. In one restaurant outbreak, chickens were cut into portions on a chopping block and cooked thoroughly by boiling. Unfortunately the boiled chickens were prepared for table on the same block, and the dinners in the restaurant got food poisoning. Oddly enough the germ was called *Salmonella blockley*.

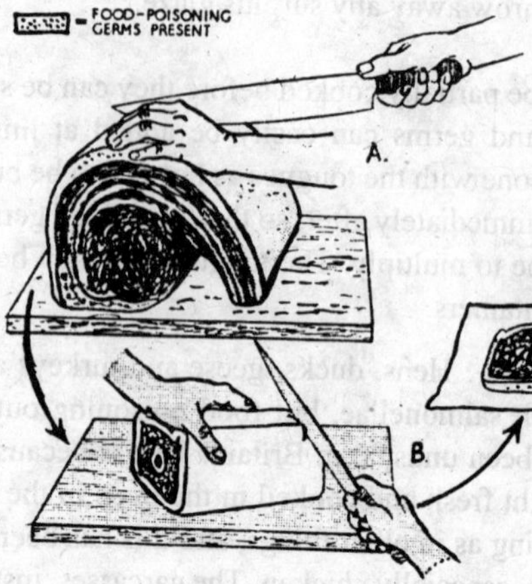

Fig. 9.2 : Germs can spread from raw meat : (a) to cooked food, (b) they multiply on the latter, (c) and give the customer food poisoning.

Butchers' shops : In a butcher's shop, if germs are present on one piece of raw meat, they may easily be spread to other pieces. They can be carried on the butcher's hands; or the counters, the chopping-blocks, the knives and the scales or anything else in the shop, including the refrigerator, may be contaminated and so spread the germs. However, with the exception of sausages, the meat will be contaminated only on the surface, and cooking will destroy the germs.

Butchers mostly sell cooked meats as also raw meat; this includes tongue or corned meat sold loose in the shop, pies and all kinds of browns and jellied meats. If these are touched by hands that have handled raw meat, cut with the same knives, weighed on the same scales, stored in the same refrigerator, or displayed for sale on the same counter, any germs present on the raw meat or on the equipment will be transferred to the food and will multiply on it till eaten by unsuspecting customers. Once again the rule applies — *raw and cooked meat should not be handled together.*

Canning Factories

Most canning processes produce a sterile food inside: any bacteria present are killed. Canned food is therefore very safe food indeed, possibly the safest of all. A can is a very sturdy metal box which can take a good deal of buffeting. Very rarely a minute fault develops in a can, and then a drop of water may be sucked inside, and the a drop of water may contain dangerous germs. The meat or other food in the can provides ideal conditions for multiplication and when the can is eventually opened, though the food still appears sound, it is teeming with germs. This is nearly always the story when canned food causes disease—accidental contamination of the food through a minute leak in the can after sterilization—but it must be emphasized that such incidents are rare. It has been calculated that there is less than one outbreak of food poisoning for every 400 million cans of food consumed. The risk is very small, but when such an outbreak occurs, it can be devastating.

All the above applies to canning methods which aim at sterilization by subjecting the food to very high temperatures. Sometimes this high temperature method is thought to make the meat less attractive, and manufacturers aim only at 'pasteurizing' the meat. This means that not all germs are killed, though most disease germs are. The temperature in the middle of a large can may be very little, if at all, above the temperature needed to kill food-poisoning germs so that the safety margin is small. This method is used mainly for large tins of pork, veal and tongue and these meats are first treated with preserving salts such as nitrates, nitrites and sodium chloride. This pickling followed by pasteurization achieves a fairly high degree of safety: but it not sterilization, and food-poisoning germs can survive the process.

9.5 The End of the Trail

Bakehouse, Kitchen and the Home

The trail is long. Germs spread from one animal to another; the journey from farm to farm, then on to market or abattoir, and from there at last to the bakery, kitchen and the home—this is a trail that may take years to cover, but salmonellae are great wanderers and well fitted to stand the buffeting on the way. To some extent they are helped along by man's ineptitude, carelessness or ignorance, but it is only at the end of the food-poisoning trail that these factors become a major menace. This is the menace we must now consider.

A Germ Looks at a Kitchen

For a germ looking for a place to settle down in, a bakehouse or kitchen has many attractions. Apart from one or two very hot spots which must be avoided, the place is pleasantly warm. There are one or two very dry areas, but mostly it is a moist warmth that pervades the atmosphere. Many surfaces are hard and metallic and give no lodgement for ease-loving organisms, but there is mostly some softer wooden table, bench or ledge with hair-breadth chinks and cracks into which a wandering germ can retire and raise a family. The air of a kitchen tends to be a little dusty, but the dust contains many nice food particles, and some eventually fall into the chosen chinks and cracks and ensure a food supply for the resident germs. A food-poisoning germ in such a situation has a long and fruitful life ahead of it.

Entry of Germs

In case of salmonellae, we have seen that they may already be present on the surface of raw meat or inside raw sausages. They may also be introduced by a member of the kitchen staff who happens to be a salmonella carrier. Such a person has probably had an attack of diarrhoea sometime in the past, and, though now perfectly fit, continues to pass the germ in his faeces. It is all too easy to contaminate the hands at toilet and salmonellae may thereby get on the fingers and be transferred in the kitchen to food or cooking equipment. Workers in kitchens and bakehouses may be infected by the contaminated food that is passing through their hands, and it is not uncommon, in an outbreak of food-poisoning, to find that several of the employees are excreting the germ in their faeces.

The germs of dysentery, the shigellae, are parasites of man only, so that if a food-borne outbreaks of dysentery occurs there must be a human carrier among the kitchen staff. He or she has probably had an attack of diarrhoea a short time ago, and is usually easy to detect by bacteriological testing.

Clostridium, welchii is a common germ. It is present in dust and soil, and may easily find its way into kitchens and bake-houses, on boots or packing cases, or on sacks of potatoes or vegetables. It is generally present in the intestines of man and animals, thus that it may find its way on to raw meat or the fingers of people handling it. In a contaminated kitchen it may be present almost anywhere, so that if an outbreak of *Clostridium welchii* food poisoning occurs, it is almost impossible to pin down the source of entry of the germ.

Staphylococci are among the most common of germs and some staphylococci can be found on the skin of nearly everyone. Fortunately most of these are non-pathogenic and cannot cause disease, but some are pathogenic, though most of these are still not of the food-poisoning variety. The most common sites are the nose and the hands. Generally there are no signs at all on the nose or hands of the carrier, and the staphylococci are found only when swabs are taken to the laboratory. Sometimes the infected worker has a sore on the fingers or on the nostrils — a small septic spot or boil is the most common finding. When this is so, he or she can shed millions of germs into the surroundings and into the food he or she is preparing.

Germs all over the place : When germs get into a place where food is prepared they can spread in an astonishing way, and it is very easy to grow them from tables, floors, window-ledges, taps, bowls, dish-cloth, napkins and brushes, as well as from food utensils. It is not easy to keep them out, but one rule which follows from what has been said above is: *Workers who have any skin sores or have recently had diarrhoea must not be allowed to work where food is being prepared*

9.6 The Manipulation of Food

Germs don't jump into food. They have got to be placed there. This involves manipulation or handling of food. Manipulation comes from the Latin word, *manus*, the hand, so it means the same as handling, i.e. touching or moving food by hand. The chef or baker can contaminate food directly with the germs from his own hands, by handling the food on some contaminated surface or in some contaminated bowl or basin, or by placing it against some previously contaminated food. The raw materials are present in abundance—meat, lard, eggs, flour—and germs can easily be added to them. When the two come together, the raw materials and the germs, how do the involved manipulations of cooking affect the outcome?

Roasting, Boiling, Frying, Stewing

When fresh food is roasted, boiled, fried or stewed for adequate periods,.the cooked food is always safe to eat. *Food poisoning cannot be caused by fresh, fully-cooked food eaten hot from the stove*. The important word are 'fresh', 'fully-cooked' and 'eaten hot'. If any of the three are left out the guarantee of safety no longer holds. If the food is deep-frozen, inadequate thawing may prevent heat penetration; if the food is only lightly cooked, the temperature may not rise high enough to kill pathogenic germs; if the food is allowed to cool for hours before being eaten, germs may get in and multiply in the tepid food.

Safety in cooked food is the product of temperature and time. The temperature which matters is the temperature in the centre. Most food-stuffs are poor conductors of heat. This means that it takes a long time for the heat on the outside to penetrate the inside.

For a pie or joint of 6 lb. weight a gas-oven mark 4 to 6 might be chosen. This corresponds to an oven temperature of 176° C. to 204° C. (350° F. to 400° F.). This is the temperature applied to the outside of the joint or pie, but it would take at least 2½ hours for the temperature in the centre to reach 100° C. (212° F. or boiling-point). The common cooking rule of 20 minutes to the pound and 20 over is barely sufficient, but 30 minutes to the pound is adequate. Of course, food-poisoning germs are killed at temperatures well below boiling-point, and in any case these organisms cannot be present inside a fresh, solid joint, though there might be some inside a pie.

Underdone or rare? Rolled joints are a different problem, for in these the outside of the meat is rolled into the inside and food-poisoning germs can be rolled in too: but if time and temperature are correct, they are all killed. Many people like their roast beef or steaks rare or under-done. The colour of cooking meat turns from red to brown at about 70° C. (158° F.), and red beef may be well below this, perhaps just a little over 60° C. (140° F.). It takes up to ½ hours at 60° C. to kill food-poisoning germs, so there is not much margin.

Light Cooking

Cooking may be 'light' because of low temperature, short time, or both. Cooking fat melts and reaches a temperature of 149° C. (300° F.) before it begins to smoke and burns. If sausages are cooked in such fat, left in long enough and turned till brown on both sides, any salmonellae in the centre are killed. If they are fried at a lower temperature or for too short a time, the temperature in the centre will not be high enough, and we have already seen that salmonellae may survive gentle frying inside a sausage for 15 minutes. A steak well fried is sterile, but if the centre is rare it may not be: the temperature of the rare bit may be just marginally above 60° C. (140° F.) and if the time is less than 30 minutes, salmonellae could survive could survive the ordeal. A hard fried egg is safe, for egg coagulates at 72° C. (161° F.), well above the lethal point for food-poisoning germs: but a soft fried egg could still contain live salmonellae. With hen eggs this is most unlikely, for though salmonellae can be present in the shell they are not present inside: but duck eggs may have salmonellae inside them. The danger from hen eggs arises mainly from bulk liquid or dried egg, not from shell eggs.

Most people like an omelette to be moist in the centre. This means that the egg albumen has not coagulated so that the temperature is below 72° C. (161° F.): as the cooking time is very short, food-poisoning germs could escape. This might be a danger with omelettes made from duck eggs. Baked egg custard is usually cooked at oven gas mark 4 or 176° C. (350° F.) for about ½ hour: if the custard sets properly, the internal temperature must be at least 72° C. (161° F.) and any salmonella present will be killed: but if the custard is only half cooked, germs could survive, and outbreaks of food poisoning have been caused in this way. In pouring custard, the temperature must be kept below boiling after the eggs have been added to the milk and as the cooking time is short, food-poisoning germs might survive. However, if the custards are made from freshly-broken shell hen eggs there is no danger, for salmonellae on the shells, if they got into the eggs while being broken, would not have a chance to multiply. If the eggs are beaten and kept sometime before cooking, or if unpasteurized bulk liquid or dried egg is used, the danger is much greater.

Junket is made from warm milk curdled by the action of rennin. The temperature must be kept carefully around body temperature, 37° C. (98° F.), because above that temperature the rennin is destroyed and will not act on the milk. This temperature is ideal for germ multiplication, but if pasteurized milk is used (and no other *should* be used) there is no danger at all, and junket is a thoroughly wholesome food. Sometimes eggs are added to junket: when this is done, germs could be introduced and would thrive rather than be killed at the temperature and time of junket making.

The most dangerous form of light cooking is where the heat is adequate but time too short and this is especially the case with the roasting of chicken and other birds. The difference between a solid lump of fresh meat and a bird is that food-poisoning germs are present only on the surface of the meat, whereas there is every chance that they are present deep in the carcass of a bird. They get there during evisceration if salmonellae are present in the bird's intestine, or the carcass may be contaminated in its journey through the poultry plant. It is therefore essential that heat penetrates to the centre of the bird during cooking. It is recommended in some cookery books that ½ to 1 hour's roasting at gas mark 4, 176° C. (350° F.), is sufficient. This may produce nice browning of the outside, but it will not sterilize the inside of the chicken. At least 2 hours at this temperature is essential for a chicken and much longer is needed for big birds. It is most important with deep-frozen birds that they are thoroughly thawed before cooking, otherwise the heat in the centre will be far below sterilizing temperature, and serious outbreaks of food poisoning have occurred in this way.

Slow Cooling

Most foods are poor conductors of heat. They take in heat slowly, but they also give it off slowly. A joint of roast beef, left at room temperature, takes many hours to cool down in the centre, and this applies to pies, tarts, custards and all other cooked foods. At first the temperature falls quickly. (If food is left for only a few minutes before being brought to the table we complain that it is 'cold'.) Later the loss is much slower as the temperature gradient between the food and the surroundings becomes less, and from 50° C. to 37° C. (122° F. to 98.6° F.) it is very slow indeed. Now this is just the temperature range which food-poisoning germs like. Salmonellae revel in such an environment, as do staphylococci and *Clostridium welchii*, and these last two also produce their toxins in food in this temperature range. If a few germs in the food have survived the cooking or if there are germs in the atmosphere of the kitchen, on tables or utensils, in floating dust, or on the hands of food handlers and these germs get into the cooling food, they find conditions ideal and start to multiply at an alarming rate. A bacteriologist, with all his laboratory know-how, cannot produce a medium that germs like so much as a pie, or a joint or a custard trifle, cooling slowly in a warm kitchen. Remember a germ divides in 20 minutes in good conditions. During a night in a pie, one germ can multiply to several millions.

Staphylococci and *Clostridium welchii* both produce toxin when they multiply, and these toxins, especially staphylococcal toxin, may not be destroyed by boiling. Staphylococci grow readily in almost any kind of food, but they like especially meat, trifles, custards, cakes, pies, or cold potatoes. Often this kind of food is allowed to cool slowly, and this is what staphylococci enjoy. They multiply recklessly and produce lots of toxin. The food is probably eaten cold next day, and is bound to cause trouble: but even if it is re–heated, the toxin will not be destroyed.

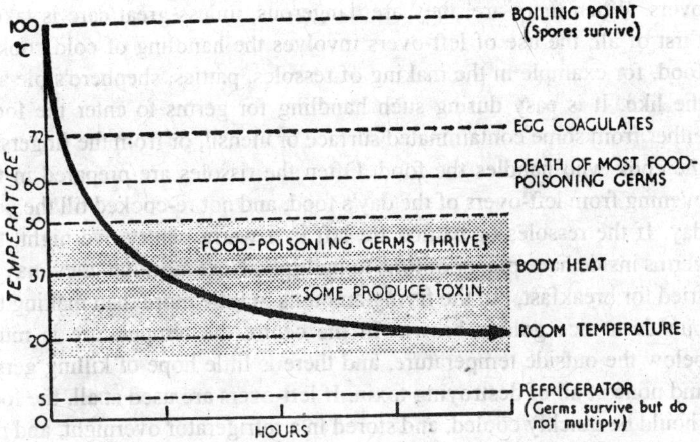

Fig. 9.3 : Food temperature graph. Hot food cools quickly at first then very slowly through the range that suits of food-poisoning germs:

Clostridium welchii is more fastidious than staphylococci. It dislikes air and oxygen and will not grow on the surface of a cream cake or trifle: it prefers the condition of low oxygen tension which it finds inside a stew or pie. Even very thorough cooking does not kill clostridial spores, and if a few of these are present in a slowly cooling stew or pie, they germinate at around 50° C. (122° F.) and the vegetative germs then multiply quickly and produce toxin. Ideal conditions are provided when a stew is kept warm in a bain-marie or when cooked meat is sliced, covered with gravy and kept warm on a hot plate. Such food after a few hours can be teeming with germs and also full of toxin, and the two together cause food poisoning. Often this food is eaten the same day without further heating, but even if the food is left to cool overnight and served the next day, the reheating is usually anything but thorough, and both germs and toxin survive. If meat, or stew or pies are not eaten immediately after cooking, they must either be kept at a temperature well above 60° C. (140° F.) till served, or cooled very quickly and kept in refrigerator till required for reheating. The re-heating next day should be thorough, but in large establishments it is much safer to have a rule that all food is eaten the day it is cooked. *Slow cooling of hot food is more dangerous than any other kitchen procedure.*

Left-overs

If the above given instructions are followed, there can, be no left-overs. When there are, they are dangerous, unless great care is taken. First of all, the use of left-overs involves the handling of cold cooked food, for example in the making of ressoles, patties, shepherd's pie and the like. It is easy during such handling for germs to enter the food, either from some contaminated surface or utensil, or from the fingers of the cook who handles the food. Often the rissoles are prepared in the evening from left-overs of the day's food, and not re-cooked till the next day. If the ressoles or patties are left in a warm pantry overnight, the germs inside have plenty of time to multiply. Next morning, they may be fried for breakfast, but the frying amounts to little more than making the outside a nice golden brown. In the centre the temperature is much below the outside temperature, and there is little hope of killing germs and none at all of destroying toxin. If left-overs are used at all, the food should be quickly cooled, and stored in a refrigerator overnight, and not shaped or handled until just before cooking. And the cooking must be thorough, not just gentle warming. Failure to observe these rules can result in an outbreak of food poisoning or dysentery.

Cold Buffet

There are few sights more attractive to the eye of a hungry man (or woman) than an elaborate cold buffet. Golden brown chicken cutlets, cold sausages, boiled hams, galantines, potato salads, meat pies, stuffed pastries, creams, cakes, gateaux and trifles—all tantalizingly attractive, but all foods that require a good deal of handling in their preparation. For convenience, the buffet table has often to be laid and furnished several hours before the meal and this means that the food is left for those hours at room temperature—and the room is mostly very warm. Unless the food has been prepared with scrupulous care, germs may be present in it, and they are given time to multiply in it while the food lies on the buffet table. Many outbreaks of food poisoning have followed cold buffet meals. The safeguards are obvious from what has been said above: the food must be quickly cooled after cooking, and stored overnight in a refrigerator, and kept there till the latest possible moment before the buffet meal.

9.7 ***Keeping Food Safe***

In this chapter we have dealt with the hazards of food preparation in the bakehouse and kitchen. If all the precautions are observed, the final product should be wholesome and harmless, but there is still the danger that the food can be contaminated *after* careful preparation. The greatest danger is when raw and cooked food are handled in the same area or by the same worker. A chopping-block or preparation table is certain to be contaminated by any germs present on the surface of the raw food handled on it. If cooked food is handled on the same surface, it will pick up some of those germs and all the care taken in the cooking of the food will be wasted. It would seem to be common sense not to run this risk, yet this is probably one of the commonest faults in kitchen routine, and food poisoning has been caused in this way. Any utensil or container, a knife or a jug for example, used for raw food is contaminated by it. It is madness to use them for cooked food, but this happens commonly.

It is all so easy to appreciate these facts in theory, but it is not so easy to realize them when at work in a clean kitchen. The important word in that sentence is 'clean'. 'Clean' does not mean the same as 'safe', for germs are invisible: one cannot see them on a clean knife, a clean jug, or a clean slicing machine. One cannot see them on a clean table or in the clean air of the kitchen, but if one remembers that invisible germs may be present on raw food and are only too ready to spread to cooked food, one can take steps to make this difficult for them. *Cooked food should never be handled in the same area as raw food: utensils used for raw food must be sterilized before being used for cooked food.*

10

The Menu in a Restaurant

10.1 Introduction

A restaurant menu makes more demands on its readers than any other form of literature. There is a time limit for reading a menu: Although its scope is vast, it must be comprehended quickly, often under the eye of a supervisor who knows its contents by heart.

So much for the strains and hesitations. Menu reading has its rewarding side as well, and there is probably no other literary form which can excite the imagination and the senses so strongly. The tempting smells, tastes and colours of the dishes are called up, and, in addition, the imagination paints still-lifes of pantry tables laden with fish, joints and game, vegetables and fruits, spices, vinegars and oils, creams, sugars and sponges, as well as pictures of ships and planes conveying these from the corners of the earth, and armies of chefs, confectioners and pastrycooks eager to fashion them into masterpieces of the culinary art.

Such is, or ought to be, the effect of the menu on the customer. He should have no reservation about the quality or attractiveness of the dishes it describes. The food hygienist looks at the menu from a different angle; he is concerned with the wholesomeness and safety of the food provided.

10.2 Contents of a Menu

Snacks : These are snacks. They activate the appetite for what comes after and so must be light and delicate. They stimulate the mouth and the stomach. Then there are anchovies and pieces of ham, slices of mortadella and salami, fish, eggs, eels, olives, asparagus, peppers, chervil and parsley. It all makes quite a show, but it involves a good deal of handling behind the scenes. Most of the constituents of a tray are very safe foods. Pastry and toast do not encourage bacterial life, and anchovies, herrings, peppers and olives are too salty, too acid or too oily to allow bacteria to multiply. Hams, sausages and fillings carry some risks, for they may be contaminated during handling, and, as they lie on the trolley for several hours during meal service, germs have time to multiply on them before the last dinners are fed. These foods must therefore be prepared with scrupulous care, separately from any raw meat, and kept cool till required. They are really a form of cold buffet.

Soups and Broths : Soups and broths are boiled slowly for hours, and are sterile when served hot. One can scarcely eat a safer food. Yet a broth is very little different from the fluid media used in laboratories for coaxing germs to grow. In fact, bacteriologists call their media broth, but they do not use it boiling hot, for that would kill their precious germs. Herein lies the only danger of soups and broths. They should not be allowed to cool slowly and then be left tepid for hours, for if germs get into the pot they will grow rapidly in such an ideal medium. If such a broth is reheated, vegetative germs might be killed but toxins might not be destroyed, and outbreaks of food poisoning may be caused.

Cold consomme is safe only if it has been cooled quickly and kept in a refrigerator, but if there is any carelessness in its preparation, it can become a consomme of germs. Ideally soups and broths should be consumed on the day they are cooked, without any cooling and reheating. Stock-pots should be looked on with suspicion: if tolerated at all, they should be maintained near boiling-point and kept carefully covered. Otherwise they are dangerous.

Fish : The diner has little to fear from fish. Food-poisoning germs do not inhabit the intestines of fish, for fish are cold-blooded animals, and salmonellae and clostridia like a bit of warmth in their habitats. In some waters the germs of botulism get into fish, and in fermenting fish dishes the germ may multiply and cause botulism. But such dishes are served in tents and wigwams, not in restaurants, and the dinner in the restaurants may order any fish on the menu with complete confidence.

Shellfish in a restaurant is almost certainly safe, for there are regulations covering its gathering and cleansing. Mussels, oysters and cockles are gathered in shallow inshore waters which are generally polluted by sewage, and salmonellae may be present in such waters.

Shellfish suck great quantities of water through their bodies, filtering minute food particles from it for their nourishment. Unfortunately they also filter out any germs that are present. And build up a high concentration in their body tissues. People have bathed in such water without coming to any harm, but have caught typhoid fever from eating shellfish gathered there. Just as shellfish become infected from filtering polluted water so they cleanse themselves when put in tanks of clean water, and this is the basis of purification techniques used for them. If shellfish are on the menu, it can be assumed they have no stains on their character.

When fish is left over and compounded with potatoes into patties or fish cakes, it loses its good name and acquires all the disadvantages of left-over cookery. Salmonellae or staphylococci may be added from the hands of food handlers or from any contaminated dust or debris in a substandard kitchen. Outbreaks of food poisoning or even dysentery have been caused in this way, but it is the handling, not the fish, that is to blame.

The Main Course

Meat : The roast beef can stand up to any safety tests. It is well browned on the outside, and even if the centre is still rare, no germs can get through its tough exterior, provided it is not a rolled rib roast. Served hot from the oven it can be offered to friend or foe, Englishman or foreigner, without a thought for any intestinal repercussions. It is staunch and sterile. The same cannot always be said for the gravy. If this is prepared straight from the roasting pan it can be poured with confidence over a plate of roast beef and Yorkshire pudding, and can be a noble addition to the national ensemble. But sometimes gravies come by devious routes from dubious stock-pots and pass through lingering temperatures on the way. Such infusions are not germ proof.

What is true of roast beef is true of grilled steaks. They are sterile. They may be tough, but, from the bacteriological point of view, they are fit to eat. Stews are not so safe: one wants to know something of their history. If they are cooked on the day they are eaten and served hot from the pot they cannot contain germs; but if the stew has been allowed to cool or kept just warm for several hours before being brought to the table, anaerobic organisms like *Clostridium welchii* may have got into it and multiplied, and the stew may be full of germs and toxins. The same is true of any beef, pork, mutton or ham prepared long before the meal, covered with gravy, and kept warm for hours on a hot plate or in a bain-marie.

Ragouts and banquettes are normally made from fresh meat, veal, chicken or mutton, and if the cooking is thorough and continuous, they are wholesome and sterile. If there is any interruption in the cooking, or worse, if they are allowed to cool and are kept for reheating next day, they can be the most dangerous of meat dishes for, when cool or tepid, they are ideal culture media for anaerobic germs. A blanquette is sometimes made from left-over beef or pork: the *veloute* or white sauce is made by mixing white stock with the roux of flour and butter, and sometimes yolks of eggs are added at the last moment to make Allemande sauce. This is poured over the meat, and the total quantity of cooking the blanquette gets may be minimal—all very well if the cold meat is sterile to start with, but not enough to kill any germs that may be present in it.

Pies and poultry : A pie consists of a bottom, sides and top, enclosing a filling of meat, mutton, poultry, game or pork. This is a prosaic view of a romantic object. Into a pie may go beef from the Argentine, mutton from Australia, wheat from Canadian prairies, spice from the Orient, lemon juice from the Mediterranean, lard from a Danish piggery and butter from an English dairy farm. The constituents are blended and baked with skill and the finished product is a work of art. It is also unfortunately a potential nest of infection. The geographical background of the pie exposes it to all the bacterial hazards of animal husbandry, transport and slaughter and even if it escapes these, contamination can occur in the bakehouse or kitchen, and salmonellae, staphylococci or clostridia can be stowed away deep in the heart of the pie. It can be made safe by heat penetration; a golden crust does not guarantee this.

Poultry flocks are generally infected with salmonellae, and these germs are present in the centre of the carcass and not, on the surface only. This means that thorough cooking is a must. Roast beef, with a rare centre, is a safe dish because there are no germs in the centre of a joint, unless it is rolled rib roast, but an underdone chicken is quite a different matter, because it is the deeper parts of the bird that are liable to be contaminated by germs in its intestines, and if the temperature does not rise high enough in the centre, the germs survive the cooking. A fresh chicken after roasting is likely to be thoroughly cooked, but with a deep-freeze bird, unless thawing is thorough the centre of the bird may be very underdone, and outbreaks of food poisoning have been caused in this way.

Game birds are usually free from dangerous germs in the wild state. Contamination may occur from faulty preparation, particularly if cooling and reheating are involved. The centre of a cold game pie is an anaerobic reserve and clostridia can flourish there with little competition.

The cold table : Cold meats are attractive both to man and microbes. The latter can share the dish only through faults in preparation. Microbes get their best chance to invade the meat after it has been cooked and while it is still cooling. They must be helped on to it, for the distances the most mobile germs can cover on their own are entirely microscopic. So the food handler must help the germs along, by transferring them from his own person or from some contaminated raw meat to the cooked food. Once in the food, and the temperature right, food-poisoning germs get a new zest for life.

Vegetables, pickles and sauces : Vegetables when cooked are very safe foods, and even raw they rarely cause trouble. Salads are often suspect, but the water they are washed in, or the dressing they are smeared with, may be more to blame than the greens, cucumbers or tomatoes. If there is any doubt these can be washed in chlorinated water and this may be advisable when fresh vegetables have to be stored a long time as, for example, on board ship.

Germs do not like acid, so pickles and sauces made with vinegar are likely to be germ free. Aigre-douce sauce has vinegar and wine in it, as has *Lyonnaise* sauce; and *bigarade* sauce is sharpened with the juice of bitter orange; but bread sauce is compounded of breadcrumbs and butter, with consomme or stock added to it, a slightly risky combination if there are faults in stock-pot hygiene. *Bechamel* is a sauce of milk, flour and butter, safe enough if made and served the same day, but a good medium for germs when tepid. *Hollandaise* sauce is pure butter and yolk of eggs, probably too rich in fat for the easy growth of germs, but if the eggs are contaminated with salmonellae, they might overcome this difficulty and multiply. *Veloute* is a basic white sauce made of white stock mixed with flour and butter: made properly, it is cooked for over an hour and must then be sterile, and if used at once is above all suspicion. It may, however, be strained through a cloth and beaten till cold, both manipulations which may cause contamination. It is then used as a base for many other sauces, and some new ingredient may add to the danger. White sauces in general are good media for germs except when cooked for a meal, and served hot with it. It may be convenient to prepare them in advance and keep them overnight, but it is not sound food hygiene to do so.

Mayonnaise and Salad Dressing : Mayonnaise is a slightly mysterious substance, both in the origin of its name and sometimes in its preparation. Some hold it should be magnonaise from manier, to stir, a likely enough derivation, for stirring is the essence of mayonnaise preparation. Others say it should be mahyonnaise from Port Mahon in Minorca, possibly because of the olive oil imported to France after the place was taken by the Duc de Richelieu in the mid-eighteenth century. A third view is that it is a corruption of Mayonnaise from the Old French *moyeu*, meaning the yolk of an egg. All three suggestions are plausible, for this sauce is certainly a combination of egg yolk and olive oil which has to be stirred to make it into mayonnaise. The only other constituent is vinegar, and of this there may be much or little. This last ingredient is the most important from the food hygienist's point of view, for if a lot is added, the mayonnaise will be too acid for germs, but if only a teaspoonful or two, they could multiply in the mixture. Made fresh from egg yolks, as it usually is, mayonnaise is a safe food: but if the yolks are separated and kept for sometime before being used, salmonellae could grow in them and would be killed when beaten into mayonnaise, for no cooking is involved. However, as far as food poisoning is concerned, the record of mayonnaise is a good one.

Salad dressing is pure oil and vinegar, and food-poisoning germs will not grow in such an oily acid mixture. Salad cream is sometimes made from the sieved yolks of boiled eggs mixed with cream and vinegar without further cooking. This is fine if the cream is safe, but unfortunately this cannot be taken for granted. Sometimes flour is added and raw egg yolk, and unless such a mixture is well acidified with vinegar it could furnish food-poisoning germs with very attractive pabulum.

SWEETS AND CHEESE

A full menu lists a wide variety of sweets. There are pastry sweets such as tarts, pies and flans, and the heavier sweets such as rice and semolina and weighty boiled puddings; meringue sweets and whipped cream sweets, ice-creams and fruit sweets; custards and caramels, sweet omelettes, crepes and fritters; cream cake or gateaux, trifles, charlottes, sponges, souffles, moulds, fools and jellies. They are the products of ingenuity and art, but also unfortunately of constituents which must be scrutinized for their liability to convey food poisoning.

CONSTITUENTS OF SWEETS

Eggs

We have already seen that poultry may be heavily infected with salmonellae, and it is not surprising that eggs become contaminated. If an egg is dry and cold, germs find it difficult to get through the shell . If eggs have been soaked in a liquid culture of salmonellae in the laboratory, but no salmonellae have been found inside the egg. If a warm moist egg is laid on cold wet ground, salmonellae may be sucked through as the egg cools, and this may explain why germs are found more often inside duck eggs than hen eggs, for ducks generally lay their eggs out in the open, while hens prefer a warm dry nest. The percentage of eggs infected inside the shell is quite low and there is rarely any danger in eating a boiled egg, but salmonellae are very often present on the shells of hen and duck eggs alike, and when eggs are broken in bulk to form commercial liquid egg, salmonellae from the shells or from the odd infected yolk get into the mix and then multiply and contaminate vast quantities of the product.

Bulk liquid or dried egg has been responsible for many outbreaks of food poisoning. When they are used in bakeries and kitchens, the liquid and powder contaminate utensils and tables, and the egg dust can form an aerosol or mist of infectious particles. Baking with egg depends on its coagulation, and this begins at 72° C. (161.6° F.), well above the death point of salmonellae which is around 60° C. (140° F.) Cakes, sponges, Swiss rolls, custard tarts and other egg products are therefore sterile when they come out of the oven, but may easily be contaminated again as they cool on a table in the air of the kitchen or bakehouse. Much more dangerous are the confectionery practices which do not involve cooking the eggs: fillings for cakes and buns often contain artificial cream or raw whipped egg, and these have caused serious outbreaks of food poisoning.

Dried egg is produced either by forcing the liquid egg as a spray into a warm chamber where it falls to the floor as a powder, or by spreading it on warm rollers where it forms a yellow film which powders readily. The egg must not be allowed to coagulate during the drying process, as this would make the dried egg useless for baking: the usual temperature in drying is below 51.5° C. (125° F.). This is well below coagulation temperature of 72° C. (161.6° F.). but it is also below the temperature lethal for salmonellae, and the egg powder is not sterile.

All bulk liquid egg must be pasteurized. The pasteurization temperature is 64.4° C. (148° F.), and the liquid egg is exposed to this temperature for 2½ minutes. This would not cause coagulation of whole liquid egg nor of separated yolk, which begins to coagulate at about 71° C. (160° F.). It is therefore very desirable that whole egg or egg yolk should be pasteurized before drying.

Separated egg white (known when dried as crystal albumen) begins to coagulate around 63.3 C. (146° F.), which is below the temperature used for pasteurizing bulk liquid egg and such pasteurization would spoil the product. A lower temperature of 57.2° C. to 57.8° C. (135° F. to 136° F.) for a longer time may be effective, for salmonellae are killed at 55° C in an hour Egg white is difficult to pasteurize after drying, but exposure to a temperature of 54.4°C (130° F) for days on end has been tried and may be effective. Although conditions have been greatly improved there must still be some uncertainty about the safety of some egg products. The key to success is to find some way of pasteurizing whole or separated eggs before drying which will not affect the baking properties of the dried product.

Milk

Germs are very fond of milk. The germ of tuberculosis used to be very common in cattle in Britain and many cases of bone tuberculosis in children were caused through drinking infected milk. Nowadays as a result of strict veterinary control of herds, tuberculosis is a rare disease in cattle, and milk-borne tuberculosis is now almost unknown. Typhoid fever has often been caused by drinking raw milk. Cases still occur. Food-poisoning germs such as slamonellae or staphylococci may get into milk and multiply in it, and, in the past, even outbreaks of scarlet fever and diphtheria have been spread by milk.

There are two things essential in producing safe milk: these are clean production and heat treatment. Clean production alone will not produce safe milk, for germs do not mind whether milk is clean or dirty, so long as they can get into it; but, of course, if production methods are scrupulously clean, germs have less chance of getting in. Heat treatment by pasteurization destroys all disease germs, and it is the only way of guaranteeing that milk is safe. Milk may be sold as *Untreated* only if it comes from cows that have been proved by tests to be free of tuberculosis: to get a licence to produce such milk, the dairy farmer must maintain very strict standards of cleanliness and herd hygiene. Such milk used to be sold as *Tuberculin Tested* milk: it is very clean milk and there are no tuberculosis germs in it. Other germs can get in, inspite of all the precautions the farmer can take. Such milk reaches a high standard of purity, but it is absolutely safe *only if it is pasteurized.*

Pasteurized milk can be contaminated after pasteurization. The cap of the bottle may be fouled and some germs may get into the bottle as it is opened. Most contamination occurs after the bottle is opened and the milk poured into other containers. There is nothing magic about pasteurized milk: the heat kills any disease germs present in the milk before heating, but germs that get in later can flourish in the pasteurized milk. Milk should therefore be handled with great care, covered from dust stored in a refrigerator.

Dried milk. Dried milk powder is liquid milk from which the moisture is driven off by the spray or roller method. The temperature is too low to kill all germs, but milk for drying is first pasteurized and the drying process then results in a safe milk. The milk powder is produced and packed under very careful factory conditions, but germs may get into the powder after the tin or packet is opened. They cannot multiply because the powder is too dry, but when water is added to reconstitute the milk the germs can multiply in it if it is left at room temperature. Obviously milk should not be reconstituted till required for cooking: if it must be made earlier, it should be stored in a refrigerator.

Evaporated and condensed milks. These are milks from which part of the moisture has been removed by heating under a vacuum. Only pasteurized milk is used and so these milks when canned are safe. Once the can is opened, germs can enter. In evaporated milk they can multiply quickly, unless it is kept at refrigerator temperature, but in sweetened condensed milk the high sugar content makes the milk almost self-satirlizing and bacteria cannot multiply in it.

Thus milk can be an ideal medium for disease germs to multiply in, and milk can therefore spread disease: but if milk is produced under good hygienic conditions, heat-treated and then kept carefully it is a safe nutritious food.

Cream

Cream contains most of the fat of milk, and, as a medium for germs to grow in, it is very similar to milk. Cream can be separated from the milk after pasteurization, or the cream itself can be pasteurized before or after bottling, so that cream should be as safe as milk. A great deal of contamination may take place after the cream is pasteurized, for its is often sent to retailers in cans, and bottled or put in cartons in the shop. The cream that eventually reaches the bakery or the kitchen may be far from sterile. As cream is handled a lot in the kitchen or bakery, and often added to cakes or trifles long before they are eaten, any germs present have plenty of time to multiply, and food poisoning may follow.

Artificial cream. The term 'artificial cream' covers a wide variety of products and preparations. At one end artificial cream may contain all the protein, fat and carbohydrate of dairy cream and, at the other, be compounded of water, starches, vegetable oil and other substances which never issued from a cow's udder. It can therefore be a wholesome and delicious addition to a ravishing sweet; on the other hand, it may be only a germ-contaminated emulsion squeezed over some soggy cake or sponge. It is better to keep the term 'artificial cream' for products that are made from milk constituents, for example dried milk and butter, and to use the term 'synthetic cream' for those that contain none, but this is not the normal practice.

All artificial cream offered for sale must be clearly labelled 'artificial cream'. More important than a label are the contents and preparation of the cream. A good product is one made from dried milk, butter and sugar, and pasteurized in the factory. Such a cream can be as high in food value as fresh dairy cream, and as far as germ contamination is concerned, it is often much safer, for the standard of hygiene in artificial cream factories is very high, and a can of branded artificial cream is a thoroughly satisfactory product. Other artificial creams are made from sugar, fat and eggs, and here the danger is from contaminated egg: if bulk liquid egg is used it must be pasteurized and is safe, but imported dried egg might not be.

Cream fillings. Various butter creams are made by confectioners and cooks in their own bakeries and kitchens. These may be made from sugar syrup, egg yolk and butter, or from egg yolks, fresh cream, sugar and butter and flavoured in many ways. Praline cream is butter cream with almond powder beaten into it, mocha cream is flavoured with coffee, while Plombieres cream has liqueur added at the last moment. All these creams involve some degree of heating, but there is a good deal of whipping and beating during cooling, and the heating would not be enough to destroy germs that were present at the beginning or got in during these processes. Cold cream fillings can be made by beating icing sugar into fresh dairy cream, or by mixing sugar with egg white and butter.

Marshmallow and gelatin. A rich cake filling is made from mixing marshmallow with egg white and adding it to beaten butter and milk powder. Marshmallow itself is made from dissolved agar-agar or gelatin mixed with egg whites and sugar. Gelatin is manufactured from animal bone or hide which may well be contaminated with salmonellae: these are likely to be destroyed by heat during manufacture, but recontamination may occur later when the gel has cooled. Agar-agar is not an animal product but is derived from seaweed: it is not contaminated at source, but if germs do get on it, they grow readily.

All these fillings contain materials that may be contaminated with germs at the onset—egg, dairy cream or gelatin. The greatest danger lies, however, in contamination during and after preparation: the air may contain germs in dust. Tables and other working surfaces may have a coating of germs, but above all, plant and utensils may become contaminated by one batch of material and, if these are not thoroughly cleansed, germs can multiply in any film of cream left in them, and grossly contaminate future batches. All the handling, of course, lays these creams open to contamination from a human source, and staphylococci from confectioner's nose or fingers have caused outbreaks of food poisoning in this way.

Cornflour filling. Synthetic cake filling can be made with water, cornflour, margarine or cooking fat, egg yolk and sugar. This is a nutritious food though it has no claim to be considered as cream. The cornflour is thoroughly cooked, the fat or margarine is added at 65° C. (130° F.), still above the lethal point for food-poisoning germs, but the egg yolk is added when the mixture has cooled to 45° C. (130° F.) which is well below the lethal point. The main danger, however, is in contamination after preparation, for this filling is a good medium for germs.

Butter, Margarine and Fat

Germs find it difficult to digest fat. A few uncommon germs do produce an enzyme known as lipase which breaks down fat into simpler substances which they can then use as food, but this does not apply to any of the food-poisoning germs. Very occasionally the germ of brucellosis of Malta fever has been found in butter and, more rarely, the germs of typhoid fever and tuberculosis; but these are epidemiological curiosities, not examples of common methods of spread. Moreover, butter is now made almost always from pasteurized milk, so that even these minute dangers are removed.

There is evidence that some fats and fatty acids are germicidal, that is they destroy germs, and it is true that germs find life in a fatty medium very uncomfortable and cannot survive in it for long. This does not mean that one can be careless, for if butter or fat were contaminated at a late stage of preparation a few germs might survive and cause infection; but apart from these risks, butter and fat can be considered as safe and wholesome foods.

Margarine is prepared by blending fat free milk with vegetable oils derived mainly from nuts—groundnuts, coconuts, and palm kernels are the most common. Sometimes whale oil is added to the blend. Every effort is made in the manufacture to reproduce the physical properties of butter. Spreading factor can in fact be controlled in margarine more easily than in butter, which is often too soft in summer and too hard in winter. The nutritional values the fats of margarine are equal to those of butter, and when vitamins are added to margarine there is no difference between the food values of the two The only factor that eludes the manufacturer is the subtle flavour of butter. As regards freedom from germs, margarine is even safer than butter, for most of its constituents are of vegetable not animal origin and it is manufactured under conditions of maximum hygiene. However, the prejudice against margarine persists, and people consider butter as a natural, but margarine as an artificial food. It is difficult to understand how butter can be considered as a more natural food for man than margarine, because butter is prepared entirely from cow's milk, that is intended by nature to be sucked from the cow's udder by its calf, whereas margarine is derived mainly from the fruits of the earth.

Flour

Flour contains 70 to 75 per cent carbohydrates, 8 to 12 per cent protein, less than 1 per cent fat, traces of cellulose and mineral salts and about 14 per cent moisture. It is much too dry for germs, and bacterial food poisoning has never been caused by flour. Chemical food poisoning has occurred on several occasions due to contamination with dangerous chemicals in transit or storage.

Cornflour is obtained from maize. Like flour it is dry and unsuitable for germ life. When heated with water or milk to over 75° C. (167° F.) it gelatinizes, and as it cools, it sets as jelly. Such a jelly is sterile but, if it became contaminated later, germs could grow readily on it.

Rice flour is derived from the rice plant. It is often used as a dusting powder on confectionery and cakes. It forms a jelly in the same way as cornflour. It is safe when dry, but a good medium for germs when moist.

Arrowroot is not a cereal but is obtained from the thick root, or rhizome, of several plants. The story that American Indians used root flour to heal arrow wounds seems to be untrue: the Indian word for root flour is araruta and has nothing to do with English arrows. Arrowroot is used to make milk puddings, moulds and blancmanges. It is also used, by boiling with water and sugar, to make jelly for pouring over flans and tarts. It is sterile when prepared, but may be contaminated after it cools.

Sugar, Honey and Jam

Most germ need some of the simpler sugars like glucose or lactose for food, but only in weak solutions. They cannot survive in strong concentrations. A strong solution of a substance like sugar or salt always draws fluid from weaker solutions. Germs are composed mainly of water with low concentrations of salt and sugars in them (as indeed the human body is), and if germs are placed in a strong solution of sugar or salt, water is drawn out of their bodies and they die of thirst or dehydration. This is the principle used in the preservation of food by salts and sugar.

Jam and honey contain very high concentrations of sugar in solution and germs cannot survive in such conditions. Sometimes a mould grows on the top of a jar of honey or jam, but this is because there has been some condensation of water at the top of the jar that lowers the concentration of sugar in the surface layer—low enough for moulds and yeasts, but still much too high for other germs. Sweetened condensed milk is another product with such high sugar concentration that germs cannot live in it. Sugar and all preparations containing strong solutions of it are, as far as germ contamination is concerned, very safe foods.

Cocoa and Chocolate

Cocoa is a powder obtained by roasting and crushing cocoa beans: these grow in large pods in cocoa trees. The powder contains a substance quite akin to the caffeine of tea and coffee and has a stimulating effect on the brain: this substance is a stimulant drug, not a food. The powder also contains cocoa-butter which is a food with a high calorie value. A cup of cocoa therefore has food value of its own, whereas a cup of coffee or tea has none apart from that supplied by the added milk and sugar, but, of course, there are only one or two teaspoons of cocoa in a cup so that the amount of nutriment is quite small.

In the crushing and roasting process a good deal of cocoa-butter is extracted from the beans though some still remains in the cocoa powder. To make chocolate, more cocoa-butter is added to the cocoa powder and the mixture is blended with sugar. (Confectioners need a very pliable chocolate and this is produced by adding still more cocoa-butter to the blend.) Plain chocolate contains about 30 per cent fat and 60 per cent sugar. Milk chocolate has dried or condensed milk added to it: this raises the fat content and lowers the sugar content. From the safety point of view, the concentrations of fat and sugar in all chocolate, as well as the heating roasting and blending, guarantee a germfree product.

Coconut : Coconut, for confectionery use is derived from the dried fleshy part of the fruit of the coco-palm. It contains up to 50 per cent fat and, being of vegetable origin, should be free of any germs derived from animals. Unfortunately the standard of hygiene practised in the harvesting and drying of the coconuts in tropical countries has been in the past deplorably low, and the dried coconut has been found contaminated with salmonellae of human and animal origin. These germs probably find it hard to multiply in the desiccated coconut, but when this is scattered on some sponge, gateau or trifle the germs find the moisture they need there and multiply quickly. Outbreaks of food poisoning and paratyphoid fever have been caused in this way. Fortunately the standards of hygiene in the producing countries have been raised and the danger of contamination greatly reduced. Importers in Britain also subject the coconut to heat sterilization, so that coconut is now usually a safe product.

The story of coconut food poisoning is important, because it emphasizes how every process in food production must be supervized and scrutinized. Time, money, legislation and care are required to produce a safe end-product, yet all may be undone at the last moment by a few flakes of nuts contaminated some thousands of miles across the ocean.

10.3 The Sweets Trolley

The first thing the food hygienist notes is that many of the articles on the trolley contain materials on which germs grow readily: there is plenty of carbohydrate and protein and plenty of moisture. The second point he notes is that the trolley is wheeled into the restaurant before the beginning of the meal and remains there for several hours, so that there is also plenty of warmth and time. What he cannot see is whether germs are present or not; this he can only judge if he knows what has gone into each article and under what conditions it has been prepared.

Gateaux, Eclairs and Sponges

All these have been prepared from flour, eggs, milk and sugar, and, fresh from the oven, are free from food-poisoning germs. There are three dangers thereafter. The first is that some form of artificial cream or other filling has been added and this may be contaminated with germs. The filling moistens the dry sponge or batter and the germs may then multiply there in great comfort. The second danger is that the air and dust of the kitchen or bakehouse may be full of germs and these may alight on the food after baking. The third danger, one common to all foods, is that a human carrier may contaminate the food after baking. In all three cases, the germs, once on these foods, will find the conditions very suitable for multiplication.

Trifles and Meringues

A trifle is the same thing, from a germ's point of view, as a gateau or sponge: a trifle is moisture and so easier to grow on. Trifle left-over is an ideal medium for germs. At room temperature, a few staphylococci could multiply to millions overnight: in a refrigerator the numbers would not increase. Alcohol is an excellent disinfectant, but the amount of sherry in a trifle tends to be meagre and would not discourage determined germs.

Meringues are made from white of egg and sugar. The more egg white used, the lighter is the meringue. They may be used as fillings, coverings or dried as a shell. They may be made in the cold, or baked in the oven. If the whites of fresh shell eggs are used there is very little danger, but bulked egg white (crystal albumen) is not always free from germs and there is therefore some risk in the cold process. Meringue shells are produced by baking at 121° C. (250° F.), a temperature which dries the meringue but does not caramelize the sugar: it would also kill any germs present in the mix. The only danger from shell meringues lies in the filling: if it is contaminated, the meringue is no longer a safe confection.

Charlottes, Fools, Moussses, Jellies and Blancmange

The factor common to all these sweets is that they are largely uncooked. Charlotte russe, for example, is prepared by beating cold custard with gelatin and whipped cream: the eggs in the custard, the gelatin and the cream are all liable to be contaminated with germs (Apple charlotte is a very different sweet composed of bread, butter, apple and sugar and thoroughly cooked in the oven.) Fools are cold fruit purees whipped with sugar and cream, and desert mousses have a base of chantilly cream (double cream plus sugar and vanilla) with chocolate or other flavouring or fruit puree added. In both fools and mousses the cream may be suspect. The basis of jellies is gelatin. This is dissolved in a little cold water, then boiling water and the flavourings are added and the jelly left to set. If salmonellae were present in the gelatin, this degree of heating would hardly kill them. Blancmange originally was prepared from pounded sweet and bitter almonds set in a mould with sugar, water and isinglass. Today blancmange usually consists of a cold cornflour pudding variously *flavoured*. It is a rather insipid dish, but germs, if they got into the cold mould, could certainly flourish in it.

Custards and Caramels

Baked custards and caramel creams depend for setting on the coagulation of egg. This requires a temperature of at least 72° C. (161.6° F.) and this is well above the temperature required to kill food-poisoning germs. These custards and creams are therefore safe when first prepared: if left to cool and stored before being served, germs might gain entry and would multiply rapidly.

Pies, Tarts and Flans

In Britain a tart has two crusts and a flan only one. In America an English tart would be called a pie, and an English flan a tart. Names apart, they are the same things and consist of pastry with fruit filling. Most fruits are acid and germs do not multiply in acid. Moreover, the more acid the environment the more quickly are germs destroyed by heat. Fruit tarts and flans are therefore very unlikely to harbour food-poisoning germs when they come out of the oven. A jelly is usually added to flans after they are baked: this may be made of gelatin or of cornflour or arrowroot and there is some risk of germs being added then, but, even so, they are unlikely to multiply in the acid medium. A more likely danger lies in the whipped cream (natural or artificial) which may be used to decorate a flan, or the cream which is served to eat along with the tart.

Ice-cream

Ice-cream is an oil in water emulsion. The oil phase consists of fat and the water phase contains milk solids, sugar and *flavouring*. This oil-in-water emulsion is called the 'mix'. In Britain the mix must contain 5 per cent fat and 7½ per cent milk solids. If the ice-cream is to be sold as 'dairy ice-cream', the fat must be milk fat: but in 'ordinary' ice-cream this is not the case and the fat is usually vegetable fat. To improve the stability and texture of the ice-cream, emulsifying and stabilizing agents are added. The most common are glyceryl monostearate and gelatin. These keep the texture firm but not too hard, and avoid too rapid melting and the formation of large crystals. The result is an even, velvety, icy substance which melts smoothly in the mouth. It is of high calorie value and, though sold in millions of tons for its cooling properties, it must, when digested, add to the production of body heat. At the moment of consumption it is accepted eagerly as a refreshing frozen sweet, without a thought for calories. Unfortunately before it is frozen, it is highly acceptable food for germs. Unless precautions are taken, ice-cream can cause disease and there have been large outbreaks of food poisoning, typhoid and paratyphoid fever caused in this way.

Heat treatment. Under the existing Ice-cream (Heat Treatment, etc.) Regulations ice-cream mix must be heated and must conform to other time and temperature requirements. The principles are those applicable to the safe handling of any food—heat to kill germs, quick cooling to prevent germs entering and multiplying at intermediate temperatures, and cool storage till required. The rules allow three methods of pasteurizing the mix. It may be heated and held (a) at 65.6° C. (150° F.) for 30 minutes, (b) at 71.4° C (160° F.) for 10 minutes or (c) at 79.4° C. (175° F.) for 15 seconds. After methods (a) and (b), the mix must be cooled within 1½ hours to 7.2° C. (45° F.), and it must be kept at or below this temperature till it is frozen. The temperature after freezing must not rise above — 2.2° C. (28° F.): if it does, the mix must be re-treated. After method (c) the mix may be transferred, with aseptic precautions, into sterile, air-tight containers, and the law does not require that it be kept at a specified low temperature before freezing.

Soft ice-cream. These pasteurized or sterilized mixes are called 'soft ice-cream' and are supplied as such to restaurants and retailers who are responsible for freezing it. A mix may be evaporated after heat treatment and sold as cold mix powder: after it is reconstituted it must be frozen within 1 hour.

Safety

Ice-cream mix is a good medium for germ growth. The most likely source of infection is a human carrier of staphylococci or of typhoid or paratyphoid germs. These germs, if they are given time and warmth, multiply rapidly in the mix and the staphylococci produce toxin. Subsequent freezing destroys neither the germs nor the staphylococcal toxin. Before the heat – treatment was made compulsory, ice-cream was a dangerous article of food. With the heat treatment, rapid cooling and freezing required by law it is now one of the safest. The dangers that remain arise at two sources: (a) in the manufacturer's plant and (b) in the handling of the unfrozen mix after it has left the factory.

Defects in the plant arise from incomplete cleaning and sterilizing somewhere along the processing line. If this takes place in the cooler or vats after pasteurization, contamination of the mix takes place, and although the temperature might be too low to permit much multiplication, enough germs, particularly typhoid germs, might persist to cause serious trouble. A pasteurization plant is not a self-sterilizing unit: it must be maintained and serviced by an engineer who understands not only the mechanical aspects but also the hygienic work the plant is needed to do.

A modern ice-cream factory is generally run on very satisfactory lines and high standards of bacteriological purity are achieved. This is true also of most restaurants and retailers, but here the human factor becomes important. Unless all utensils used for storing the soft ice-cream, for reconstituting cold-mix powder and for final freezing of the mix are cleaned thoroughly and sterilized between uses, contamination may take place. If there is any failure to observe temperature requirements, such contamination could have serious results. To handle ice-cream safely a retailer requires intelligence and some knowledge of hygienic requirements. These qualities are not always prominent in some branches of the trade, and supervision by public health inspectors is necessary. As stated above, ice-cream properly prepared and distributed is one of our safest foods, but germs are always ready to take advantage of human carelessness or ignorance.

Fruit

Fruit is another safe food, for most fruits are too acid for germs to thrive on. If the skin of fruit is dusty it is common sense to wash it, but dipping fruit, especially grapes, in a bowl of water on the table before eating it is of no value. If germs are present on the skin some would be washed into the water which would then become a source of further contamination. Fresh fruit is wholesome food and can be eaten without worrying about food poisoning.

Canned fruit has never caused bacterial food poisoning: germs are readily killed in an acid medium by the heat of the canning process. Dried fruit may be contaminated with food-poisoning germs if the standard of hygiene is low: the germs cannot multiply in the dry, acid fruit, but if this is left to soak in water, any germs present would readily multiply. No outbreaks due to this cause have been reported, probably because the dried fruit requires thorough cooking.

If one has any qualms about a batch of fruit, it can be washed in water containing 60 to 80 parts per million chlorine: this will destroy any food-poisoning germs present.

CHEESE AND BISCUITS

Cheese

Cheese is highly concentrated food made from milk: 1 pint of milk forms about 2 oz. of cheese. To make cheese from milk, a culture of lactic acid bacteria (the starter culture) is added to the milk: these bacteria attack the lactose of the milk (milk sugar) and convert it into lactic acid which causes souring of the milk. The milk is now warmed and rennet is added. This causes a curd to form and this curd is the future cheese. The liquid or whey is allowed to drain off and later the curd is pressed to squeeze out as much whey as possible. The cheese is then left to mature for several months. Cream cheese is made in the same way, except that it is not pressed and so has a more open texture and retains more moisture. Moulds can grow in some cheeses and help to mature them. Stilton, for example, is a mould-matured cheese.

Safety

Two points can be made at once. (1) If cheese is made from pasteurized milk it is safe. All commercially produced cheese in Britain is made from pasteurized milk and this is true of many foreign countries too. (2) The mature, hard cheese is much too dry to support the growth of germs.

If cheese is made from raw milk, germs in the milk may grow in the cheese. In fact, such germs can displace the starter germs. Staphylococci are often present in raw milk and cheese made from such milk may be teeming with staphylococci, and outbreaks of food poisoning have been caused in this way. Cream cheeses made from raw milk are especially dangerous, for the extra moisture in the finished cheese allows germs to multiply. Germs other than the common food-poisoning germs may be present in such cream cheeses, for example brucellae, the germs of undulant and Malta fever. (All milk in Malta is now pasteurized and one can eat the delicious Malta cream cheeses with confidence and enjoyment.) The main danger comes from home–made cheese, for contamination with germs from the cheese-maker can readily occur. Commercially prepared cheese can normally be considered as wholesome and safe.

Biscuits

As far as food poisoning is concerned, biscuits are blameless. A biscuit contains only 5 per cent water. This crisp dryness is for the diner a biscuit's main attraction, but germs will have none of it. They will put up with a good deal in their efforts to survive, but not with this degree of dryness. Our food hygienist may therefore munch his fill of biscuits, confident that whatever risks he may have taken with earlier items in the menu, he runs none at all with the biscuits.

Thus the diner can do nothing to fend off hidden dangers, for the safety of the food he eats depends on the history of its production, distribution and preparation before it is brought to the table. Most meals are thoroughly safe and wholesome, and food poisoning hits the headlines and radio and TV news only because it is an uncommon occurrence, and therefore is news. It should, of course, never occur at all. It is always due to some fault in the production, distribution or preparation of food, and these faults can be avoided. That, is why we have examined them. Only if one understands how food poisoning occurs can one hope to prevent its occurring again and again. A little learning may be a dangerous thing, but, in the practice of food hygiene, a little knowledge makes for safety.

11

The Preservation of Food

11.1 Introduction

Raw food is never a sterile substance. Meat, flour, milk, sugar, anything and everything we eat — all are natural products, all part of the cycle of nature, all subject to corruption and decay. Much of the breaking-down processes are accomplished by germs, and these germs are on our food when we get it. They cause spoilage of food, and spoilage is simply the first stage in breaking complex into simple substances, and returning these simple substances into circulation in nature. These spoilage germs are carrying out an essential job, but from our point of view they are detrimental because, they spoil our food. It becomes sticky or slimy or develops an off flavour. In other words, it goes bad.

With one of two exceptions we do not eat bad food. The exceptions are mainly game dishes though strange fish dishes are eaten in other parts of the world. We usually refer to our game as being 'high', not 'bad'. 'High' is a gastro-nomic term, 'bad' a food hygienist's term, but they both mean the same thing.

11.2 Spoilage Germs

Spoilage of food is an economic and not a medical problem, for it is not bad food which causes food poisoning, but food which appears sound. The germs which cause food poisoning are quite distinct from spoilage germs. These latter lead different lives. They are on the whole hardier than food-poisoning germs, because they have a job to do and must be able to get on with it in adverse conditions. They can often withstand heat that kills food-poisoning germs, and they sometimes multiply at low temperatures at which other germs cannot grow. The food-hygienist is concerned with both groups of germs. He must always remember that though he may kill or subdue the one by methods of food preservation, the other can survive many of his strategies and, unless controlled, can undo all his efforts.

The most obvious example is the pasteurization of milk. The temperature and time are adjusted to kill all the pathogenic germs likely to be found in milk: but these will not kill thermophilic or heat-loving germs. If the pasteurized milk were allowed to cool slowly, these germs would multiply rapidly as the milk passed through their favourite temperature range, and quickly spoil the milk. This is why it is required that milk must be quickly cooled after pasteurization. There may also be in milk some germs which can grow in the cold and they may grow even when the milk is kept in a refrigerator. This explains why pasteurized milk does not keep indefinitely, for these cold-growing germs give an unpleasant taste to the milk though they do not sour it.

The term preservation is used to cover all methods of keeping food, including sterilization as also pasteurization, curing and radiation. Much of the information has already been given in earlier chapters but is brought together here for convenience.

11.3 Refrigeration and Chilling

The purpose of a domestic refrigerator is to stop germs from multiplying on food. The germs are not killed, and will start to grow again when the food is taken out into a warm room. The temperatures below which food-poisoning germs stop multiplying are as follows:

Salmonellae	5° C. (41° F.)
Staphylococci	10° C. (52° F.)
Clostridium welchii	6.5° C. (43.7° F.)
Clostridium botulinum	10° C. (52° F.)

All these are well above freezing-point, so that there is no need to have the refrigerator at freezing-point, and a good deal of gas and electricity is wasted trying to keep the refrigerator at zero. A temperature of 5° C. (41° F.) is obviously low enough, though to be on the side one can aim at 3°– 4° C. (37.4°-39.2° F.).

Some spoilage germs can still grow at these low temperatures, though slowly, and so food cannot be kept indefinitely in a refrigerator. A lot depends on the nature of the food and the degree of contamination when the food is placed in the refrigerator, but one week is probably the longest that food should be left in a refrigerator, and four to five days is safer.

The Use of a Refrigerator

A refrigerator must be used with care and intelligence. It is meant for keeping food cold, not for cooling hot food. If hot food is placed in a refrigerator, the temperature is bound to rise and this may encourage the growth of spoilage or even food-poisoning germs. Moreover, hot food, in cooling, gives off water vapour and this can condense on other foods and in this way, too, favour the growth of spoilage germs. A refrigerator must not be packed with food, for this prevents the circulation of cold air and leads to pockets of stagnant, warmer air. Bad ventilation and dampness inside a refrigerator are signs of unintelligent use. They are also conditions that encourage the growth of moulds and yeasts.

Defrosting must be done regularly, for the mechanism cannot work properly if the cooling coils are covered with a thick layer of ice. The inside of a refrigerator must be kept clean: it is good practice to wash the walls and trays during defrosting, making sure the interior is thoroughly dry before switching on again. Some moulds and yeasts grow even on the cold bare walls of a refrigerator; washing helps to get rid of them.

11.4 The Cold Room

A cold room is really a large refrigerator. It is, however, often used for cooling joints and other foods, and because of its size, there is not the same objection to this as in a refrigerator. Because of the increased moisture, ventilation becomes very important and a fan is advisable to keep air moving and prevent pools of dank, still air. Large institutions and factories often have special cooling rooms in which the hot food is exposed to a continuous current of cold air, and rapid cooling is achieved before the food is placed in the cold room or refrigerator.

Food kept in a refrigerator or cold room is chilled, not frozen. Microbial life is slowed down, but the food itself is not changed in any way, and its length of life is limited. A refrigerator is of great value in the temporary preservation of food, but its limitations must be understood.

11.5 Deep-freezing

A deep-freezer is a refrigerator for keeping food in the frozen state; all the fluid in the food, be it a pie, a trifle or a carcass, is changed to ice. In a domestic deep-freezer, the freezing and the storing are done in the same refrigerator, but in a food factory or large institution the food is frozen before being put into the deep-freeze. This is most effectively done by blast freezing. The food is placed in a special chamber, the blast freezer, and subjected to a continuous blast of air at under –20° C. (– 4° F.). The time and temperature are varied according to the type and texture of the food, but ½ to 1½ hours is an average time at a temperature between —20° C. and —40° C. (—4° F. and —40° F.). This rapid method does not affect the flavour or the texture of the food. Slower methods can affect both.

When the food is frozen, it is packed in the deep-freeze and kept at around —18° C. (0° F.). This, of course, is far below the temperature at which food-poisoning organisms stop multiplying, but freezing does not kill them, nor does it destroy the toxins present in the food before freezing. When the food is taken out of the deep-freeze and thawed, any food poisoning germs in the food before freezing will start to multiply again, and unless the food is thoroughly cooked, food poisoning may follow. If staphylococcal enterotoxin were present even thorugh cooking would not destroy it.

Not all life comes to an end in the deep-freeze because some bacteria, moulds and yeasts can multiply at this low temperature. This is particularly the case if the freezing has not been thorough and somewhere in the frozen food there are pockets of unfrozen moisture, for moisture favours growth. Such growth of germs can lead to spoilage of the food, although serious spoilage is uncommon in efficient freezing plants.

More important than spoilage in the deep-freeze is loss of flavour; this seems to depend on a factor of time which varies for different foods, but there is very little loss till after 6 weeks. The reason for the loss of flavour is not known but it is probably due to some subtle chemical changes.

When food is taken out of the deep-freeze, it must be thawed thoroughly and cooked immediately. Thawing of a chicken or turkey takes between 8 and 12 hours. If thawing is incomplete, the heat of cooking may not penetrate the whole carcass, and any germs still alive in the centre when cooking begins are still alive when the bird is 'done'. Food taken out of the deep-freeze and thawed should never be refrozen, for germs may have multiplied during the unfreezing and the food may be badly contaminated when returned to the freezer.

11.6 Frozen Meals

The tendency has been to think of deep-freezing as a method of preserving only fresh food, or limited variety of cooked meat and fish. There is little doubt that the deep-freeze will be used more and more for cooked foods of all kinds on a commercial and institutional scale. Whole meals can be cooked and frozen, stored in the deep-freeze, and thawed, reheated and served on demand. If the period of storage is not too long, there is no loss of flavour and the food is as good as freshly cooked food. Research is still required into the mechanics of thawing and heat penetration, but with food carefully prepared and thoroughly cooked before being frozen there is far risk of food poisoning than with raw deep-frozen food. The method has already been successfully tried in hospitals, and it is bound to spread to the home. Prejudice, of course, will have to be overcome. There are always those who claim that nothing can ever be so good as home-cooked food, as if every housewife was by natural endowment a cordon bleu cook. The same type of person objects to pasteurized milk because it has been subjected to moderate heat, though he will gladly drink a glass of piping hot milk as a night cap, because he is used to it. The emphasis in this book has been on food hygiene as a discipline for preventing food poisoning, but the food hygienist is concerned with all aspects of food production. He has little use for prejudices and must be ready at all times to examine critically the methods of conventional catering. Deep-frozen meals *will* become more common. The food hygienist must be alert to meet the changes.

11.7 Heat Treatment

Heat is the most effective weapon for destroying germs and so preserving food. If the thermal death point of a germ is known, then the application of heat above that temperature to the food, provided it penetrates the food, will destroy the germ. A temperature of 60° C. (140° F.) for 20 minutes will destroy salmonellae but it would not kill *Clostridium botulinum* or many of the thermophilic (heat-loving) spoilage germs. Complete sterilization of food can be achieved if sufficient heat is applied, but the taste and texture of the food may be changed in the process, and the changes may not be acceptable. Corned beef is cooked when sterilized in the can and that is how we like to eat it, but sterilized milk has a 'cooked' *flavour* which many people dislike. Some meats, such as hams and tongues, lose texture and substance if heated too much in the can and for them less intense heat is used in the canning industry. Let us look at the main methods of preservation by heat.

11.8 Pateurization

Pasteur was the great French biologist who did much of the pioneer work in the science of bacteriology. One of his early interests was in the fermentation of wine and beer. These sometimes go sour through the production of acetic acid by germs, and Pasteur found that heating the bottled wine and beer to between 50° C. and 60° C. (122°F. and 140°F.) for a few minutes, souring was prevented. This discovery was a great boon to the wine trade and the process became known as 'pasteurization'. It was soon found that the same treatment applied to milk delayed its souring, and pasteurization of milk first became popular as a means of prolonging its keeping life. It was later discovered that by controlling the time and temperature of the process all the disease germs present in milk could be killed. This included the germ of tuberculosis, *Mycobacterium tuberculosis,* and, as a result of pasteurization, tuberculosis of the spine and bones, which had been a great scourge of childhood, almost ceased to exist. Pasteurization has for long been regarded as a heat process applied only to milk, but the term is now in use for other products such as liquid egg, ice-cream and some varieties of canned meats. It is also coming into use for the treatment of food by radiation, where heat is not used at all: food can, in fact, be treated by radiation while still frozen. Pasteurization is not sterilization, for it does not kill thermophilic (heat-loving) spoilage germs. As applied to milk, liquid egg and ice-cream, pasteurization does mean that the product is free of living disease germs. Applied to some canned foods, where penetration is not so easy as in milk, pasteurization may mean only that germs have been reduced to safe level. Where disease germs are concerned no level is really safe, and such 'pasteurized' products require to be watched carefully by the food hygienist.

Milk, eggs and ice-cream. Pasteurization of milk may be achieved in one of two ways:

(a) Milk is heated to between 62.8° C. and 65.6° C. (145° F. to 150° F.) and held at that temperature for 30 minutes.

(b) Milk is heated at 71.7° C. (161° F.) for 15 seconds.

The milk must be immediately cooled to 10° C. (50° F.). Both these types of milk must be labelled 'Pasteurized Milk'. A third method is to heat milk at 132° C. (270° F.) for 1 second in an approved apparatus which diverts any milk which does not reach this temperature. Such milk must then be sealed in sterile containers and may then be sold as 'Ultra Heat Treated Milk'. A fourth method is to filter and homogenize milk and then heat it in sealed bottles to at least 100° C. (212° F.): the time required is controlled by laboratory tests. This milk is sold as 'Sterilized Milk'. It should be noted that cream is not covered by the regulations and much contaminated cream still reaches the consumer.

The temperatures and times for the pasteurization of liquid egg and ice-cream have already been studied.

Pasteurized canned meats. Milk, liquid egg and ice-cream, pasteurized according to the regulations, are free of living disease germs. The same cannot be guaranteed for pasteurized canned hams and some other meats. Full sterilization heat treatment apparently causes shrinkage, loss of *flavour* and change of *colour* and texture in some meat products, and manufacturers wish to avoid this by subjecting these meats to a less severe heat treatment which is called in the trade 'pasteurization'. Many of these meats are also treated with salts so that are partially cured and this helps to keep down the growth of germs, but failure of heat penetration is the difficulty in such compromise processes, and these 'pasteurized' hams and meats are not sterile and may even contain disease germs such as staphylococci or salmonellae. The heat and curing processes should ensure that these are present in small numbers only, but if only a few are present after pasteurization they can multiply inside the cans if the temperature suits them. All such canned meats must be regarded as perishable and they should be stored in the cold, and this should be stated on the label. They are not suitable for long storage.

11.9 Heat Sterilisation

Sterilisation by heat requires the application of heat under pressure. The temperature usually achieved in the canning industry for this purpose is 121° C. (249° F.). The germ which it is essential to kill is *Clostridium botulinum*. The vegetative form of the germ is killed at 80° C. (176° F) but the spores require exposure to 120° C. for 5 minutes. The toxins of *Clostridium botulinum* are all destroyed at temperatures well below 100° C.

11.10 Drying, Curing and Smoking

Drying – Germs require moisture for multiplication. If the moisture content of food is reduced, germs cannot continue to live on it. This is the principle underlying the preservation of food by drying, but it is important to realize that germs are not killed by drying, and that they will grow again when moisture is added. We have already seen that this is especially true of dried egg which has in the past caused many outbreaks of salmonella food poisoning.

The oldest method of drying was by exposing food to the sun. This is still used in some countries for drying fruit and fish. One can see acres of grapes drying in the sun in Greece near Corinth: the word current, or dried grape, is probably a corruption of Corinth. In Scotland till very recently fish was salted and dried in the sun, and, in Switzerland, Bindenfleisch is still prepared by drying the meat in the open air. In most commercial production, some form of artificial heat is used. In tunnel drying the food passes through a tunnel while exposed to a blast of hot air. In belt drying food is carried on conveyor belts through hot-air chambers. In spray drying liquid is forced into a chamber where it meets a blast of hot air and falls to the ground as a powder, and in roller drying liquid is sprayed on hot roller and the dry film it forms is removed by scrapers. Spray and roller drying are both used for liquid egg or milk and the temperature may be less than 51.5° C. (125° F.), well below the temperature needed to kill food-poisoning germs. As long as dried food remains dry germs will not multiply in it, but the quantity of contamination in the dried food depends on the amount in the food before drying.

The advantages of drying are that it is cheap, and storage is made easier because the bulk and weight of the food is reduced when the water is dried out of it. Disadvantages are that the *flavour* of the food is altered, the colour is changed due to oxidation and caramelization, and, when reconstituted with water, the final product is different in shape, *colour* and taste from the original food. Nevertheless, many dried foods are very acceptable, though different, and drying is likely to continue for long as a convenient method of food storage and preservation. Vacuum drying and freeze drying are newer methods in which the drying temperature is lower and there is less change in the food. Freeze drying is the only method which produces a sterile product.

Curing

In curing, chemicals are added to the food to kill germs in it or stop them multiplying. Salt is the most common chemical used. At a concentration of 18 per cent, strong brine, it will stop all multiplication and kill many germs by osmosis, and it is really a form of dehydration or drying as far as its effect on germs is concerned. Staphylococci are very tolerant of salt, and can grow at concentrations below 18 per cent, but most other food-poisoning germs are checked at 10 per cent or even lower. The vegetative forms of *Clostridium botulinum* are not checked at under 10 per cent, but the spores may be prevented from germinating at a salt concentration as low as 3.5 per cent, although this cannot be guaranteed. Nitrites are generally added to the salt in a concentration about 100 parts per million and this seems to add to the preserving action, though the mode of action is not clear. This is the kind of curing given to some meats before 'pasteurization' in cans, and in some way the curing seems to make the pasteurizing more effective than can be explained simply by the temperature. There is a good deal more of tradition and custom than of strict science in curing, and for this reason the processes must be carefully watched, for when the exact mechanism of a process is not understood, slight changes in method can have unexpected results.

Acetic acid, or vinegar, is used in pickling foods and here the acidity of the food prevents the growth of food-poisoning germs: the pH, or acid index, of pickled food is generally as low as 3-3.5, and food-poisoning germs will not grow below 4.5. Spices are often added to pickled food but they probably have no curing action though they taste as if they should: they should really be sterilized before use. Sugar is used as a preservative in jams and jellies, and concentrated sugar is one of the few human foods for which food-poisoning germs do not care.

11.12 Smoking

Smoking consists partly of heating, partly of drying and partly of curing. The heating and drying are obvious results of the smoking. The curing is more mysterious; it may come from strange pyroligneous acids in the smoke, possibly from traces of formaldehyde. Smoking adds a tang of *flavour* to the food and often tenderizes it, especially fish. It adds to its keeping qualities, but it does not sterilize it. Smoked sausages are often eaten without further cooking, but this is not a safe practice and fatal cases of botulism have been caused in this way. Smoked food, like any other cured food, is best cooked before being eaten.

11.13 Preservatives

Some chemicals may be added to some foods. Sulphur dioxide is one of the most common. It may be added only to a specified list of foods and in prescribed concentrations. Sausages are in the list, as well as fruit pulp, dried fruit, grape juice, jams, mineral water, beer and cider. It has, of course, for centuries been used in wine-making, and it is most active as a preservative in acid solutions. It is very volatile and is driven off when any liquid containing it is boiled. It may therefore be used in high concentrations in fruit-pulp intended for jam making; when the jam is on the boil the sulphur dioxide can often be smelled and can even bring tears to the eyes.

Benzoic acid is another legally allowed preservative. It is used mainly in soft drinks, cordials, fruit juices, pickles and sauces and acts best in these acid conditions. It appears to be a harmless substance and is, in fact, present naturally in some fruits and vegetables in greater quantities than is permitted by law.

Sorbic acid added to flour, cheese and marzipan is active against yeasts and moulds. It seems to be harmless, for it is changed in the body into carbon dioxide and water. Propionic acid is another substance which attacks moulds. Moulds, of course, are killed when bread is baked, but other moulds may alight on the bread after baking and grow readily. This leads to a great deal of waste bread, but if propionic acid is added before baking, it keeps down mould growth afterwards. Propionic acid is present naturally in the body, and a small amount added to flour would appear to be harmless. People often object strongly to the addition of chemicals to food and, of course, care and thought must be exercised, but all food is composed of fats, carbohydrates and proteins, and these are, of course, all chemicals.

11.14 Radiation

When a patient is X-rayed, visible rays penetrate his body and throw a shadow on a film placed behind or underneath his body. These rays consist of infinitesimally minute particles of electrically charged matter and they vibrate in a wide scatter of wave lengths. Gamma radiation is similar to X-radiation but the wave lengths are contained within a narrow band: it has similar powers of penetration. Both X-rays and gamma rays can have a damaging effect on the cells of the body, and it is for this reason that very strict precautions are observed in an X-ray department to protect workers and to limit the amount of radiation to which they are exposed. Just as the rays penetrate the body, so they can penetrate foodstuffs, and the principle behind radiation of food is to use the power of the rays to penetrate the food, while adjusting the dose of the rays so that germs are destroyed but the food itself is not damaged in any way.

Gamma radiation has already been used on a large scale in medicine for the sterilization of syringes, and also for the destruction of anthrax spores in hair. Its use in food is just getting beyond the experimental stage. It has certain advantages over other methods. It has very great powers of penetration and can be used to sterilize food after it has been canned or packeted. It can penetrate large joints or even carcasses, and bulked quantities of goods such as feeding stuffs. It produces scarcely any heat and can be used on frozen food such as poultry or frozen meat without raising the temperature more than a degree or two, if at all: in fact, radiation is possibly more effective at very low temperature and the dose can be reduced accordingly. It may be used to sterilize food completely, but sometimes a smaller dose is used to achieve only a reduction of germs. The term 'pasteurization' is used for this process just as it was in the canning of hams but the difference with radiation pasteurization is that penetration is complete, whereas with heat pasteurization of solid food penetration cannot be guaranteed. The exact action of radiation is not fully understood, but at the dosage used for foodstuffs it probably acts mainly by energizing the atoms in molecules of water in the food and these, the hydrogen and hydroxide radicals, then become highly active chemically as reducing and oxidizing agents. Direct action on organic matter is probably minimal.

The possible uses of radiation are to destroy spoilage germs and so prolong the keeping life of food: and to destroy disease germs and so make it safe to eat. It could obviously be used to sterilize animal foodstuffs and this would attack salmonella infection at one of its main sources. It could also be used to sterilize carcass meat after slaughter and especially frozen imported meat. By destroying spoilage germs it would prolong the refrigeration or deep-freeze life of food, possibly indefinitely. Applied to fresh fish it would ensure its freshness throughout shop life. Meat, vegetables and other foods already in cans or packets could be sterilized without any alteration in texture or quality: this would remove the uncertainty which clings to pasteurized canned meats. It would destroy moulds on fruit and so make transport and distribution easier. It could possibly also be used for many agricultural purposes such as to delay the sprouting of seed potatoes or to eliminate insects or other pests from grain and other foods. It might also be used to destroy worms and similar parasites in food fed to farm stock.

These are all very satisfactory features of radiation applied to food. One of the main disadvantages is expense. Another might be organization for it would hardly be possible to set up large numbers of widely scattered plants: but both these factors must be balanced against two advantages, the elimination of disease germs both in animals and man, and the saving of vast quantities of food which are lost by spoilage. More important than expense, however, is safety, and this is why radiation of food is still largely experimental. Toxic products might be induced by radiation or, far more serious, the food itself might become radioactive and its consumption be dangerous. All that can be said at the moment, as a result of experimental feeding of animals with irradiated food, is that neither of these hazards seem to occur with the doses used. Accidental overdosage might produce very grave danger. Another possible disadvantages of radiation might be its effect on the nutritive value of food, but all tests show that this is very slight, and not more than with other methods. We must, of course, not forget the germs themselves, for they do not readily give up the struggle for existence. They have already shown this by their ability to develop antibiotic resistance. There are just hints from experimental work that they may be able to develop radiation resistance. Man was caught unawares with antibiotic resistance: he must guard against this newer possibility.

Radiation of foodstuffs probably represents a break-through in food preservation, but for a long time yet we must depend on traditional method. Even if radiation becomes the traditional method, the food hygienist will still have to remain vigilant, for irradiated food, like all other processed food, can still be contaminated after treatment.

11.15 Home Preservation

The housewife cannot compete with commercial preservers of food, and there have been many cases of botulism due to the inefficient home-canning of meat, fish and vegetables. *Clostridium botulinum* is not killed even after hours at 100°C. (212°F.). This is the temperature of boiling water, and therefore any method of canning or bottling meat, fish or vegetable which relies on boiling only is dangerous: the meat is certainly cooked, but, when it cools, the conditions inside the can are anaerobic and favour the growth of *Clostridium botulinum*. If these spores germinate they produce toxin inside the can or bottle, and if the food is eaten cold, even if the cook just tastes it before heating it, botulism is certain to follow. The toxin of botulism is destroyed by heat so that if the food is thoroughly heated, it might be safe to eat, but the risk is far too great.

The use of a pressure cooker may help the housewife who is determined to do her own home-canning. A good pressure cooker exerts a pressure of 0.7 kg./sq. cm. (10 lb./sq. in.) and this allows a temperature of 115.5° C, (240° F.) to be reached inside the can. This may be just enough to kill all the spores of *Clostridium botulinum*, but a temperature of 121° C. (250° F.) is safer and this is beyond the range of the household pressure cooker.

Acid fruits are much too acid for *Clostridium botulinum* so fruit bottling is completely safe, even without a pressure cooker. The same is also true for jams and jellies where the acidity and sugar concentration together defy any germs to grow. Sometimes a little condensation water forms on top of the jam beneath the cap, and, if air gets in, a few moulds may be able to grow there, but they are not dangerous. Pickles and chutneys are equally safe because of their acidity and sometimes also their sugar content.

The chief method of preserving open to the modern housewife is the deep-freeze, and the domestic deep-freeze does not differ from the commercial, except that the housewife depends on the deep-freeze to freeze and preserve the food, whereas the commercial operator usually freezes the food before storing it in the deep-freeze.

Freezing does not improve the food in any way. The housewife takes out exactly what she puts in. If the food she puts in is contaminated, it will be contaminated when she takes it out. The deep-freeze is convenient but not magic.

12

Food Hygiene

12.1 Introduction

Food poisoning is not caused by food, but by germs in or on it. These germs are living things and, like, man they survive and multiply only in favourable conditions. We know that man dies of starvation if he cannot get food, he may freeze to death if he lacks clothes or shelter, or he may die of heat and thirst if he cannot reach water. In some respects germs are tougher than man. They can tolerate being frozen for months together and can live in dry dust for years. On the other hand, many of them die rapidly at a temperature not much above that of a hot bath or shower. We can destroy germs by making things too hot for them, but they can elude us in the cold interior of a refrigerator or in a dry dusty crack in a bakehouse or kitchen. To defeat germs, we must remember all we know about them.

We have already traced the passage of germs from the stock farm or poultry house, or even from across the sea, through abattoirs, poultry plants, food factories, shops and bakeries till they reach our kitchen and from there get on to our dining table. Now we will study how to interrupt their adventurous life following three considerations must guide our practice.

1. Food-poisoning germs are often present on food when it reaches the kitchen. This is true of raw meat which may be contaminated with salmonella or *Clostridium welchii*. Germs may have contaminated the food at any point between the farm and the kitchen: they may have reached the farm from across the sea.

Germs may be present in the kitchen itself, and can easily get on and into food. The germs may have got into the air and dust of the kitchen from some infected food and may persist there for a long time. They may be brought in on sacks of vegetables or on boots and shoes. Probably the most dangerous source of all is the food handler in the kitchen: he may be a carrier of staphylococci, salmonellae, shigellae (dysentery germs) or *clostridium welchii.*

A kitchen may be quite clean to look at but it must always be *considered as a dangerous area. Germs, remember, are invisible. All that glisters is not gold, and `clean' and `safe' do not mean the same thing.*

2. Food that is good for man is also food for germs. In a warm and moist place the germs revel and multiply in many foods, but they die in heat and stop multiplying in the cold, so they like to get into cooked food that is left for a long time in the nice, warm atmosphere of a kitchen or cupboard. They have a weakness for left-overs, pies, creams, sweets and cooked meats and they can build up deadly populations in them.

3. Germs cannot get on to food by themselves. They depend on the carelessness or ignorance of man at some stage of their journey. In the kitchen they look to the food handler to help them on their way. He often assists them unawares.

It is easy to state simple facts such as the above, but not so easy to practise what they imply. To avoid the spread of infection in animals before and after slaugher could involve great changes in animal husbandry, veterinary methods and slaughter-house practice. It would also need vast sums of money. This does not mean the problems need not be tackled. They must be, but one should know the size of a problem beforehand. As food hygienists we should be aware of these problems and press for their solution, though it is not our job to tackle them. The spread of infection in food factories, bakeries, shops and kitchens *is* our concern and it is our job to do everything possible to *avoid* it.

The most important thing is to be aware. Stainless-steels are fine in a kitchen, but if there is a thin, almost invisible film or cream inside a stainless-steel mixing bowl, it becomes a dangerous piece of equipment. A refrigerator is essential in a kitchen for keeping food cold and preventing germs multiplying, but if a food handler places contaminated food in it, he may easily get the germs on his hands and transfer them to other food in the refrigerator. These may start to multiply when the food is taken out next day. It may appear hygienic to wipe plates or glasses with a clean cloth, but the cloth, though clean, may be covered with germs. A glass display cabinet protects food against the entry of germs, but if the food is already contaminated when placed in it, the cabinet becomes an incubator for the germs, we can go on adding examples and we will have to look at more examples in the next few chapters, but we do not need to examine and assess the good and bad points of elaborate factory or kitchen equipment. Awareness is a much better protection against the spread of infection than a battery of the latest equipment. We must understand the mechanism of any equipment or machinery we use: but it is much more important to understand the mechanism of the spread of infection.

12.2 Food Hygiene: From Farm to Factory

The Young Infant

When an infant is in hospital we take great care to avoid infection. The feeds are sterilized, and everything around it that might harbour germs is disinfected regularly. It is nursed in a separate room or cubicle which is carefully sealed from outside contamination, and when the nurse leaves the cot-side, she takes precautions, with her clothes and her person, to make sure that she cannot carry germs to another infant. These measures are very successful, but, if there is the slightest slip, cross-infection can occur.

The Young Animal

Calves are just as liable as infants to catch infections. Actually, common infection of calves, called 'sours', is a quite similar disease to gastro-enteritis of babies and is caused by the same family of germs, the escherichiae. However, the conditions under which calves are kept are different from the isolation cubicle of an infant, and if these germs get into the cowshed they find it easy to spread from one animal to another.

Salmonellae too are common germs of farmyard animals. They cause disease in calves and piglets and also in adult cows and pigs. Many of the latter become permanent carriers, so that the germs which get into the cowsheds and pig sties eventually reach the slaughter-house floor, contaminating our meat and pork. Young chicks, too, are very easily infected, and the germs may contaminate whole hatcheries. When the chicks move on to the broiler farms they take the infection with them, where it can persist till the birds go to the poultry plant, and in the end thousands of dressed carcasses in the deep-freeze may harbour salmonella. The next stage is food poisoning in those who eat the poultry or the meat.

We can be easily critical of such a state of affairs, but it is not so easy to remedy it. The asepsis of an isolation cubicle cannot be attained in a cowshed, yet a good deal can be achieved once the need is realized. When farmers became aware of the advantages of producing clean milk, the standard of hygiene in milking byres rose so high that the term *milking parlours* came into use. So awareness is more important. The dangers of salmonellosis are not so obvious on the stock or poultry farm as are the dangers of germs in the milking parlour. The infection is mostly silent, and it is difficult to persuade people to spend time and money on something they cannot see. It is much more difficult to get them to spend money on something which appears to do no harm. Thus salmonellae may spread widely among chicks and adult birds without causing any illness in them or interfering with their growth, and no one knows there is anything wrong with the birds till an outbreak of food poisoning occurs in the human beings who eat them. Nevertheless if salmonellosis hits a farm in an acute form, killing off calves and older cows, the owner of the stock will do anything to prevent a recurrence.

Impracticable and idealistic though it may seem, yet the aim must be to achieve inside farm buildings and poultry houses conditions wherein germs find it difficult to spread. This involves sound building construction, the use of impermeable materials for floors and stalls, good ventilation, adequate water supply and drainage and the exclusion of vermin. It also needs informed personnel who understand what all these requirements are needed for, and who are therefore careful to see to the thorough and regular cleansing of the premises and the provision of maximum comfort for the animals. A clean cowshed may still not be free of dangerous germs, but there are likely to be far fewer than in a dirty one. The rearing of a calf is very similar to the rearing of an infant: they both thrive better when the standard of hygiene is high.

Animal Food Poisoning

Both meal and meat-and bone meal are mostly contaminated with salmonella: fish meal and vegetable products like cotton or groundnut cake, soya meal or copra are less contaminated. These proteins are mixed with carbohydrates, roughage, minerals and vitamins to make a balanced animal food, and are generally fed as pellets. In the manufacture of the pellets the mixture is heated and, if the temperature rises high enough, any salmonellae in it are killed. In some processes, however, the temperature does not reach 60°C. and will not kill the germs. It is much better practice to sterilize the materials by heat before any mixing takes place, because clouds of dust are raised in the mixing and this dust may contain germs and might easily settle on the pellets and re–contaminate them. Some countries have laws for sterilization of materials before pelleting, but in Britain and many other countries this is still not the case. It is largely a matter of economics and organization, but there can be no excuse for feeding contaminated food to animals. This is one form of food poisoning that can be avoided. There is no excuse for allowing it to continue.

Antibiotics

Antibiotics are used to a large extent in animal husbandry. Antibiotics that get added to the feed seem to help some young animals and birds to put on weight more quickly. Antibiotics are also used to prevent the spread of infection or to treat infected animals. Although in theory antibiotics should prove useful for these purposes, very often they do not work. This is true when the germ is a salmonella, either in animals or in man, yet vast quantities of antibiotics are given to try to deal with salmonella infections, generally with no effect. This might not matter, apart from the waste and the cost, but salmonellae and other intestinal germs develop resistance to antibiotics, and sometimes they can pass on this resistance to other germs. This is a highly complex matter that bacteriologists are studying very carefully, for if this kind of resistance affected dangerous germs like those that cause typhoid fever, the modern treatment of that disease with antibiotics, which is now very successful, would be made ineffective. Antibiotics do not solve the problems of the spread of infection.

Animals on the Move

The animals and birds are exposed to various hazards when they undertake their journey. They suffer from `stress' and `shrinkage' and the level of infection rises steeply. There is no doubt that the conditions under which the animals are herded and transported are responsible for this. A truck crammed with cattle for sheep, or a lorry loaded with crates of live birds is not a pleasant sight. There is scant attention to hygiene and also an element of brutality in the procedure: but man is a meat-eating animal, and to satisfy his carnivorous appetite, hundreds of thousands of other animals must be got to the abattoir or poultry plant, and there slaughtered and carved up for his food. Man is the consumer and must face these facts, but it is his duty, as well as being to his advantage, to ensure that the process is carried out as quickly and as cleanly as possible.

One of the most important factors in the spread of infection, both in animals and man, is the intensity of exposure to the germ. Thus, in human infections, people living in overcrowded conditions tend to suffer more from infections than those living in better houses. Infectious diseases are commoner in winter than in summer. This may be due to the fact that people stay indoors more in winter, and so are exposed for longer periods to one another's germs. Another factor is the standard of hygiene: infectious diseases are much commoner in underdeveloped countries with poor sanitation. All these factors, overcrowding, time and poor hygiene are equally important in the spread of infection in animals on the move. A good deal of discomfort is caused to these animals by bad timing. They may be held up in trucks in sidings, in lorries at markets, or in lairs at markets and abattoirs. These delays lead to overcrowding, to bad feeding and watering, and to gross soiling of the environment with faeces. All these factors give germs the opportunity to multiply and to spread from one animal to another.

Generally those in charge of animals have no knowledge at all of the danger of infection, but in Britain the Slaughter-house Regulations of 1967 made it an offence to keep animals in abattoir lairs for more than 72 hours before slaughter. However 72 hours is a long time if conditions are bad, and many animals may be infected before they reach the abattoir floor. From then on, the spread of infection can be more easily controlled, if only operators are aware that germs may be present and can spread through an abattoir, if given the chance. The chance is given by carelessness and ignorance. It can be prevented by awareness and care. It is desirable that an abattoir can be as clean as an operating theatre. But this is an exaggeration, for in the killing area there are bound to be blood and debris, but in well-designed abattoirs these can be quickly drained away or removed, and once the carcass is skinned, it can be hung, eviscerated, cooled and stored with such attention to hygiene that cross-contamination of one carcass with the germs of another can be avoided. In a bad abattoir, germs find life easy and spread so rapidly that knives, chopping-blocks and other equipment become contaminated, and salmonellae and other germs can be grown from swabs dipped in the abattoir drains.

Poultry Plants

A poultry plant is generally a long building with a conveyor belt constantly moving through it. At one end live birds enter, and at the other, oven-ready carcasses come out. These go either direct to retail shops or into deep-freeze at the far end of the poultry plant. The processes involved are culling, killing, bleeding; plucking, eviscerating, and cooling; trussing, weighing and packing. Culling means the removal of birds that are obviously sick or diseased: the other terms explain themselves. Now, poultry may be heavily infected with germs, especially salmonellae, without showing any sign of disease and cross-contamination of carcasses can take place in the poultry plant if the design and working conditions are poor. Some poultry plants are little more than improvised packing sheds and in these it is impossible to maintain satisfactory hygiene, and germs can pass freely from one bird carcass to another. The final product, the table bird, is bacteriologically unfit for human consumption, though this may unfortunately not be known till after it has been eaten.

The most important point in a good poultry plant is that it should be large enough to deal with the number of birds passing through it. There must be no hold-up of carcasses at any stage. The building should be well-ventilated so that dust from feathers and excreta does not accumulate, and built of impervious.

THE BAKER'S SHOP

The baker's shop differs from the butcher's shop in many a respects as regards the risk of food poisoning. The main difference is that the items the baker offers for sale is already cooked. This is both an advantage and a disadvantage: an advantage is that if the cooking has been thorough, any germs in the constituents have been killed: a disadvantage is that if germs have survived light cooking, or been added after cooking, they will multiply on the food in the shop and in the customer's home before being eaten. The other difference is that many of the baker's products are dry: bread is too dry for food poisoning germs to grow in it, as are pastries, sponges and biscuits. If cream, jam or other filling is added, however, the sponge or pastry above and below become moistened, and germs can then multiply easily. Pies are made and sold in the baker's shop and they contain enough moisture to permit any germs in their centre to grow rapidly. Eggs are used by bakers and they bring danger of contamination. Cream, both dairy and artificial, is used and forms an ideal medium for the growth of germs. The bakery workers themselves may be carriers of germs and they can very easily contaminate baked goods by handling them, especially when adding jams or cream fillings.

The Bakery or Back Shop

A bakery can be a very dusty place, with flour and egg and other powdery dust on the floor, on tables, and ledges, on clothes and hands, and suspended in the air. Such dust is inevitable during baking, but if it contains germs it may settle on food and contaminate it. As far as possible, there should be a clean room in a bakery and baked goods should be stored in the clean room till moved into the shop. Thus all efforts be done to keep down the dust. Floors should be made of smooth, impermeable material that can be easily cleaned, and tables and all working benches should have metal or plastic tops instead of wood that offers chinks and cracks for the dust to settle into. Good ventilation is needed.

Adequate facilities must be provided for cleaning and sterilizing equipment. Most artificial cream and liquid egg is now delivered sterile to bakeries, but if any batch is not, then mixing bowls and other equipment contaminated with it can pass the infection on to other mixes. It is, therefore, essential that all such equipment be sterilized after every use, and it is much safer to throw away any mix not used, than to attempt to keep it till the next day. Table tops and other surfaces should be washed thoroughly many a times a day. There must be wash-basins, liquid soap containers and paper towels conveniently sited for all workers: they should wash frequently during baking operations.

The Front Shop

When the cakes and other baked foods come into the front shop, they must be protected from contamination. All the goods should be covered, either in display counters and cabinets or under individual covers. They should be handled henceforth by no one staff material so that walls and floors can be thoroughly washed. It should be so designed that, although the flow of carcasses is never halted, the different processes are carried out in such areas that are structurally separated one from another: culling is separate from killing, killing from plucking, and plucking from eviscerating. The basic aim is that when the carcass is finally being trussed and packed, that part of the plant may be as free as possible from any contamination that might arise from the earlier processes. There must be a free supply of water, at least 3 to 5 gallons per bird, and this should be chlorinated so that there is a supply of free chlorine in the water during all wet processes. Feathers, droppings and inedible offals must be removed quickly from the main part of the plant, and adequate provision be provided for sterilizing crates and bins, preferably by steam, and for regularly washing down all immovable parts of the plant with detergents and disinfectants. A good poultry plant should be clean and sparkling, with white-tiled walls, plenty of light and well-washed floors: all fittings and equipment should be of metal, cleanly designed so that dirt and debris cannot collect in grooves and chinks. In such a poultry plant, cross contamination can be prevented. No other type of poultry plant should be tolerated.

12.3 Food Hygiene: The Food Factory

Transport

It is possible that some of the meat leaving the abattoir may be covered with germs. These germs will contaminate anything they come in contact with, and this includes containers and vans. These should therefore be specially designed so that they can be easily cleaned and disinfected. The walls of vehicles should be lined with metal or plastic, and the joints must be sealed so that no blood, fat or other liquid can seep through. The floor must also be made of impermeable material: aluminium alloys, studded to avoid slipping, are used. Where separate boxes or containers are used, these should also be made of metal or plastic and jointed securely. All vans and containers must be well ventilated, and also protected from dust and flies.

Disinfecting the Ambulance

When an infectious patient is taken to hospital, the ambulance may need to be cleaned and disinfected thereafter, depending on the type of infection: Generally thorough airing and ventilating of the ambulance suffices. A van used for carrying meat from abattoir to factory is likely to be in a much more dangerous condition as regards infection than any ambulance, for blood and other juices full of germs drip from the meat, and the careasses can swing against the walls on the journey. Germs, are on the surface of abattoir meat: in a patient they are usually well inside his body. Ideally, vans used for carrying food should carry nothing else to avoid danger of spilling dangerous substances on the food.

Hot and cold water must be available at departure points for the thorough cleaning of vehicles and containers. Hoses, detergents, brooms and buckets are all needed. There must be a planned routine and plenty of time to do the job. Moreover, the operators must be well informed so that they understand the importance of the job they are doing.

Factory Design

A food factory should be bright, spacious and gleaming. Walls should be finished with a smooth, light, impervious and washable surface and where they join ceiling and floor, the junction should be moulded or coved so that they are easily cleaned. Ceilings should be smooth and unbroken by joists or other projections. Ideally walls and ceiling should be fully insulated as this cuts down condensation, and the ceiling can then be finished with a hard gloss non-flaking paint. If not insulated, the ceiling may be finished in an absorbent plaster and painted with distemper. Where there is much steam, roof canopies are needed to draw the steam up and avoid it drifting through the factory. Floors should be made of impervious, non-slip material. There must be an sufficient supply of water and hoses, and adequate drainage. The whole building must be well lit, properly heated and efficiently ventilated.

The equipment should be so designed that it is easily cleaned, and stays clean while in use, with the least possible splashing of food particles, oil or water. All working surfaces should be smooth and impervious and all joinings watertight. There must be no chinks where dust and debris can collect. Metal, tiling or plastic should be used for all working surfaces, except chopping-blocks. For these, there is still no good substitute for wood, but the surface must be kept clean by frequent scrubbing with wire brushes and detergents, and when worn and cracked it must be planed smooth again.

Wash-basins must be provided where the work is going on. They must be used frequently and properly, and they will not be used if they are sited far from the work-bench, in another part of the factory, or only in the toilets. There must be an adequate supply of hot and cold water, preferably with a mixer valve, liquid soap dispensers that work well and are kept topped up, paper towels that tear off cleanly, and foot-operated bins that open and shut properly. There must be attractive rest rooms and canteens. Thus a food factory should be an attractive place to enter and a pleasant place to work in, but everyone must be aware of the danger from germs. Nothing in the factory or on the part of the workers must add to this danger. Germs will get on the working surfaces, the floors, the walls, or the hands of the workers. They must be removed by washing and disinfection. This must be part of the instinctive routine of the factory, for germs give no sign of their presence.

The processes carried out in a food factory consist mainly of preparing and cooking food. There is one exception–sausages. These are not cooked in the factory, and if there are germs on the surface of the meat when it reaches the factory, they will be in the centre of the sausages when they leave it. Their fate thereafter depends on how thoroughly they are fried. Pies and roasts are generally sterile after cooking, but the danger is that germs present on raw meat find their way on the cooked food, or that germs on the hands of workers in the factory may be added after cooking. This is liable to occur when glazes or jelly are added to meat or pies, but it can happen in such simple processes as slicing or packing. Hygiene after cooking is even more important than hygiene before cooking, for much of the cooked food that leaves the factory will be eaten cold, and any added germs will still be alive.

The Factory Autoclave

In a hospital autoclave or disinfector there is dirty and a clean side, completely separated by a solid wall and with no communicating doors. Dirty linen cannot get through to the clean side except through the autoclave which sterilizes it. The worker on the dirty side cannot get through at all. The same rigid precautions should apply in the food factory. Raw meat is the dirty linen, and cooked food is the sterile. The factory should be so designed that there is no connection at all between the two, and cooked foods should go out from the clean end of the factory to the shop and consumers, as free from contamination as the sterile linen goes from the autoclave to the wards and operating theatres. This is perhaps not wholly practicable, but there is no excuse for unintelligent design. There was a magnificent combined oven and refrigerator for cooking and cooling pies. It cost many thousands of pounds and its design was perfect, but it stood in the middle of the factory floor. Meat from the abattoir was wheeled in through the ever-open factory door. It was chopped and minced and finally conveyed with the dough cases to one end of the oven. From the other end a conveyor belt took the cool, cooked pies to be machine-wrapped in Cellophane, untouched by human hand: but on that journey they were exposed to the air which blew from the factory door across the chopping-tables and the mincing machines and which could gather thousands of germs from the raw abattoir meat on its way. There was no reasons at all why that superb oven should not have been built into a wall like the hospital autoclave. After all, an oven is really an autoclave. No one thought of it, because no one was aware of the danger. How often that word 'aware' comes into food hygiene!

Here we are not primarily concerned with details of structure and equipment. That is the function of the engineer and the designer, but they ought to know about food hygiene when they design their machines. Our concern is to be aware of the danger and to spead this awareness to all who work with food. This is not always easy. Many a times we find glaring errors in the hygiene of the place, which must be avoided.

CANNING

Canning is a process of rapid cooking and sterilizing of food. Other methods of food preservation, for example drying or refrigeration, aim at preventing germs from multiplying. The aim of canning is to kill germs in the food, and then to seal the sterile food so securely that no new germs can get into it while it is stored.

Acid and Non-acid Foods

All canning processes depend on heat. The quantity of heat depends on the kind of germ that has to be killed. Most heat-resistant disease germs, especially the germ of botulism, cannot grow in highly acid food. Cans of acid foods need therefore only be heated to boiling-point, 100°C. (212°F.), for this temperature will destroy all the other disease germs which might grow in acid conditions. This applies to cans of fruit, tomatoes, rhubarb, and similar foods, where the acidity is high and the pH (acid factor) less than 3.7. In less acid food (pH over 3.7) spore-bearing germs like *Clostridium botulinum* can grow, and some of them survive boiling for hours together. Cans of such food must be heated under steam pressure to a temperature of 121° C. (250°F.): a few minutes at this temperature kills all disease germs. This applies to cans of meat, fish, soups, vegetables, puddings and milk products.

Cooling and Leakage

When the cans have been heated for the correct time and at the correct temperature in the autoclave, the steam is released and cold water is run into the autoclave. This is done to make sure cans cool quickly. There are three reasons for this: (1) to avoid the food from being overcooked, (2) to avoid heat-loving (thermophillic) germs from multiplying during slow cooling, and (3) to avoid long hold-ups in the factory. (Thermophillic germs are not easily killed by heat: they multiply in warm food and spoil it, though they do not cause food poisoning). Rapid cooling of the cans causes rapid changes of pressure. When the cans are very hot, the pressure inside, due to steam and hor air, may be as high as 20lb. per square inch: the steam in the autoclave exerts a pressure almost equal to this, but when the autoclave pressure is released at the end of sterilization there is a much higher pressure inside the can than outside it, and this may push on any weak part of the can and cause a microscopic leak. As the can cools, a vacuum develops inside it, and this negative pressure pulls the weak part back again and closes the microleak, but as it does so it may suck in a minute drop of water, perhaps only 1/500th of 1 millilitre or less. If the water is pure, this does not matter. If it contains disease germs, they may multiply in the food with disastrous consequences. There are two obvious ways of avoiding this: one is to use only pure water for cooling, chlorinated if required; the other is to introduce compressed air into the autoclave when the steam is released and so prevent the violent differences in pressure.

Handling and Labelling

When the cans leave the autoclave, they go along conveyor belts to be labelled and packed. This may cause a good deal of bumping and battering, and this may cause a tiny leak on the steam of a can. If the can, the conveyor belt and all the other apparatus are dry and clean this may not matter, for germs do very poorly on clean, dry metal, but if the cans are wet, germs may easily slip through the leaking steam. It is very important, that the hands of workers handling the cans at this stage are clean, dry and free from sores, as otherwise germs could pass from hands to cans and might get inside.

The whole canning process is a highly complicated one, and much research has been carried out to make it the safest of all forms of food-preservation. It is not completely proof against micro-leakage. This means that the highest standard of hygiene is needed in the canning plant. When this is achieved, the micro-leak may not matter. Over 6,000 million cans of food are opened every year in Britain: the average number of outbreaks of food poisoning from canned food is 14 in a year.

12.4 Food Hygiene: Shops

Any modern food shop *looks* good. Pies are perfect, cream cakes catch the customer's eye, roast chickens make innocent mouths water, hams and tongues torture hungry gastric juices, and as the shopkeeper lifts a joint lovingly in his hands, breathing over it almost with affection, nothing can seem more appetizing and wholesome or further removed from anything that could possibly do one harm. It is all so attractive and fits in so well with sophistication that the modern shopper accepts it as an impeccable part of the luxury of living. Yet the cream in the cakes came from the udder of a cow which might have been diseased, the ham came from the hind leg of a pig that spent its life in a sty and ended it on an abattoir floor, the chicken came from an intensive flock that might have been alive with germs, and the butcher's joint of meat reached him only after a long and dangerous journey from the stock farm to his back shop. There are plenty of hazards in the modern food industry. The shopper should 'always have an eye to the mayne'.

THE BUTCHER'S SHOP

The butcher's raw material is meat. Most of it is good, wholesome meat, but some of it sometime is bound to be contaminated with food-poisoning germs. His equipment and his technique should be knowingly planned to avoid germs spreading in his shop. The important word is 'knowingly', for many, perhaps most butchers do not know about germs on meat.

The Chopping-block

One of the butcher's main jobs is to chop and cut meat into suitable joints and portions—a skilled job, but also one which involves a lot of handling. The chopping-block is made of wood—there is no other material that can take the blow of the chopper without ruining the blade or jarring the butcher's arm. However, the chopper as it sinks into the block can take meat and germs with it, and in the chinks and crevices the germs can multiply enormously. The block must be cleaned regularly and scrupulously and no cooked meat should ever be placed on or near it. The chopper and the butcher's hands also become covered with germs: they must be washed thoroughly and frequently.

Sausages and Mince

The butcher may sell sausages straight from the factory and over these he has little bacteriological control: he may contaminate the outside of the skins with his hands but those surface germs will be killed in the frying-pan. Often he makes his own sausages. First he minces the meat, and if there are germs on its surface he distributes them thoroughly through the mince. Then he packs the minced meat into a sausage machine which forces it out into sausage skins. We cannot blame the butcher if surface germs are now in the middle of the sausages, but he can be blamed if he does not thoroughly cleanse the mincing and sausage machine before the next batch is made. The practice is pounding sausage meat into the machine with bare fist is inexcusable—there may be salmonellae on the fist from meat already handled, or *Clostridium welchii* or *staphylococci* from his own body.

Mince is a safer food than sausages, for it is always more thoroughly cooked and the crumbs of meat, instead of being packed in a skin, are all exposed to equal heat. Of course, not every sausage contains food-poisoning germs, and thorough cooking or frying kills any germs that do get in.

The Counter

Germs on raw meat on a butcher's counter are bound to contaminate the counter. Any other meat placed on the counter will pick up some of these germs. Knives used for cutting the meat, scales used for weighting it, and the butcher's hands that carry it, must all become contaminated. It is impossible to visualize the spread of infection across the counter, for germs are invisible, but if swabs are rubbed on the counter, the scales, the knives, the joints of meat and the butcher's hands, and then stroked across plates of jelly in the laboratory, the colonies of germs that grow overnight will be easily seen by the naked eye in the morning. There will be many thousands of colonies, each made up of thousands of germs, and these may all have been derived from one piece of meat placed on the counter the morning before. Once on the move, germs can cover a surface quickly, but they always depend on help from some human. If this help is denied them, their life is short.

A butcher's counter should be made of impermeable material—plastic, metal, marble or tiling is suitable. Wood should not be tolerated, except for the chopping-block. The counter should be washed several times during the day with hot water containing detergent and disinfectant. The water should be drawn fresh from the hot-water supply each time; buckets of water kept under the counter all day become dangerously contaminated with germs. A fresh cloth or brush should be used each time and not used again till sterilized by boiling. There should be a wash-hand basin in the shop with a supply of hot and cold water, a liquid soap dispenser and a roll of paper towels. The practice of rinsing the hands in a bucket of water and drying them on the same cloth or towel all day should not be tolerated.

A butcher often sells canned meats, pies, brawns and other ready-cooked meat products. If he does, it is essential that the raw and the cooked food be kept separate. The same knife or cutting machine should never be used for raw and cooked meat, unless they are sterilized in between. The same balance should not be used, nor should the butcher handle cooked after raw meat without first thoroughly washing his hands. Ideally cooked meats and raw meats should be served from different counters and by different staff. This may all sound idealistic and extreme, but germs on raw meat are one thing: they will be killed in cooking. Germs on cooked meat are quite different: they will multiply on the cooked food and then go straight into the customer's mouths.

The Display Cabinet or Window

The butcher displays his wares, in a display cabinet or in the shop window. The main danger of this practice is that on a hot day such cabinets or windows can become almost incubators and any germs present on the meat are encouraged to multiply. This can be avoided by use of refrigerated display cabinets, or to display as little food as possible and keep reserve stocks in a chilled room or refrigerator. Pies and cooked meats should not be displayed alongside raw meat: they are better not displayed at all, unless in separate refrigerated cabinets. All display cabinets and window slabs must be kept thoroughly clean.

The Back Shop

Please always remember that a refrigerator does not kill germs. It stops them multiplying, but if there are germs on a piece of meat put in a refrigerator at night, they will still be there in the morning and they will start multiplying again if the meat is put in the warm shop window. It must also be emphasized that a butcher's back shop is just as important as, if not more important than the front shop. The customer does not see the back shop: the food hygienist should or customers, and they should be protected from germs that may be sprayed as fine droplets from the nose or mouth. Many of the cakes and sponges are excellent culture media for staphylococci, and in the course of a day several carriers of staphylococci are likely to enter the shop. There should be no fingering or touching whatever, and always glass between cake and customer. Cake tongs should be used by staff when serving.

None of this, of course, will do anything to prevent food poisoning if germs are already in the food when it comes from the back shop. The problem of window display enters here again. In a warm window a few germs in a few hours can increase to thousands, and the thousands soon to millions. One must compromise between complete safety and the need to display goods. A window may be very attractive, but its safety depends mainly on conditions in the back shop.

12.5 Food Hygiene: The Kitchen

In the kitchen we have come to the end of the infection trail. There have been many hazards on the way. Here in the kitchen, food may be rendered safe, or made more poisonous. There may be germs on the raw food coming in, and these can be transferred to the cooked food going out. The staff may add germs to the food, from their fingers, their noses or their intestines. There may be germs on the floors, on the tables and in the air, and these can settle on the food. Cooking temperatures may be too low and the time too short, and then the partially cooked food can take a long time cooling—conditions much appreciated by food-poisoning organisms. Perhaps there is not enough refrigerator or cold-room space, and germs still alive at the end of the day can multiply overnight while enjoying gentle warmth in the dark. Any or all of these hazards affect safety. How they are dealt with determines whether the food going forward to the table is sound and wholesome, or poisonous and unfit for human consumption.

Education

To make kitchens safe the most important thing is the education of the staff. If the chef or catering officer understands the mechanism of food infection, he can avoid it in his kitchen, even if it is not of the most modern design or furnished with the latest equipment. He must examine in detail every process that takes place, from the receiving of the raw materials till the finished product is on the table and, even later, till the meal is cleared away, the waste disposed of, the left-overs safely dealt with and the last dish or knife and fork thoroughly washed and put away. He must examine all these processes critically. The staff must know what they are doing and why they are doing it. If they don't understand a process they can ruin it, as the kitchen worker does who collects the plates from an efficient dish-washing machine and wipes them with a contaminated cloth.

Kitchen Design

The food hygienist should be able to advise on the principles that must underlie hygienic kitchen design. They are simple. Germs can lurk in the dust on floors, on tables and working surfaces, and in the air. They may be present on walls and window-sills, even on ceilings; as long as they stay there, they do no harm, but if they drop off in condensation water they can fall on food. Walls must therefore be of impermeable washable material, and insulated to cut down condensation: ceilings likewise, or if this is not possible, then finished in absorbent plaster. Ventilation must be efficient to suck out vapour and dust. The floor must be of non-absorbent finish, non-skid, even, and easily cleaned. Table tops and working surfaces must be of metal, plastic or any other non-absorbent, non-slintering material. There must be no cracks or chinks in the hygienic kitchen. All this adds up to one principle of kitchen design, that dust and fine debris must fine no secure lodgement anywhere in the kitchen.

A second principle is that there must be plenty of space, and the space must be functionally designed and arranged. In the reception area one needs enough room to be able to receive separately different types of foodstuffs: potatoes and other vegetables that may have earth still on them carrying *Clostridium welchii*, raw meat that may be contaminated with salmonellae, and cooked food such as hams and pies that should be free of germs and should remain so in the kitchen. The hams and pies should go straight to cold storage and should not make contact with the other foods till these have come separately through the kitchen and are cooked and ready to go to the dining table. In the preparation area also, there must be division of space and labour. Vegetables must be cleaned and prepared separately from raw meat and, if cans of food are being opened, this food must be handled well away from the raw food area, and kept away from it. This all seems simple and obvious, but neglect of this principle is one of the commonest errors in kitchen practice.

The cooking area must be well organized so that every process can be supervised. The chef or catering officer must be able to check quickly that correct temperatures and correct times are being maintained. When the food is cooked it must be taken to an area that has not been used for the preparation of raw food, and workers on the preparation side should not cross to the cooked food area. If the food is to be eaten hot, then it must be kept hot, above 65° C. (131° F.), so that germs cannot grow in it; it should not be allowed to go tepid, around 45°–50° C. (113°–122° F.), when germs will multiply rapidly in it: and, as far as possible, it should go forward to the table at oncé, and not lie in the kitchen for hours. If, however, the food is to be eaten cold, for example cakes or creams or cold meats, then it should be taken to a cooling area well away from preparation and cooking areas so that dust and debris cannot float over from the raw to the cooked food. This is an especial danger in bakeries where contaminated dust is an obvious risk, but the danger exists in any kitchen. Food-poisoning germs, are invisible. A second reason for separation is that kitchens are warm places, and cooked food will cool very slowly there, and any germs that get on it will be encouraged to multiply and produce toxin.

Finally, there must be plenty of cold storage space to avoid any hold-up in the flow of food through the kitchen. This cold storage should be structurally separated from the kitchen to prevent dust or contaminated air flowing from one to the other. The cold storage area for cooked food must be entirely separate from the chilling room used for the reception and storage of meat and other raw foods.

Washing up

Dish washing or washing up is a routine practice in every kitchen. There is no good reason why it should be. No surgeon would tolerate the washing the cleaning of instruments and towels in the middle of the operating theatre, and no chefs should be needed to have dirty dishes washed where he operates. Washing-up sinks and machines should be outside the kitchen.

The effect of heat. The aim of washing up is to remove all food particles from dishes, crockery and cutlery, and to kill any germs that may be on them. Two agents are available for this purpose, heat and chemicals. Heat acts by melting fat and grease and so helping to remove food debris: it also kills germs by direct action if it is hot enough. The hottest temperature most hands can bear is around 50° C. (122° F.). This is not hot enough to destroy germs, and if dishes are washed at this temperature, germs can still be grown from them afterwards. Usually these are not disease germs, and in any case their presence on plates and cutlery, though highly undesirable, is not a very great danger, for the germs will not multiply on dry dishes. On mixing bowls and similar utensils by contrast, the presence of germs is very dangerous, for when the bowl is next used the germs will be able to enter the mixture and multiply if given suitable temperature and time. Clearly one-sink washing is not satisfactory for such utensils. If a second sink is available, the dishes can be rinsed in this in water at higher temperature, say 80° C. (176° F.), and this will destroy disease germs. Failing this, a disinfectant or bactericidal substance can be added to the washing water. In a dish-washing machine, where hands play no part, water at higher temperature can be used, 60° C. (140° F.) for the wash and 80° C. (176° F.) for the rinse.

Soap, detergents and disinfectants. Chemicals are always added to washing water, usually one or other of many avialable detergents. The best-known detergent is soap, though it is usually not regarded as a detergent at all. It belongs to the anionic group of detergents and is the sodium or potassium salt of fatty acids. When any salt is added to water it dissociates into its ions: sodium chloride, common salt, separates into sodium and chloride ions; the sodium is the cation and the chloride the anion. Some detergents owe their action to their positively charged anions, and are called anionic detergents, some to negatively charged cations, the cationic detergents. A few have no electrical charge at all and are non-ionic. The only practical point in knowing this is that anionic detergents can be used along with soap, but the cationic detergents cannot. On the other hand some cationic detergents have bactericidal or disinfectant action as well as their detergent action. The anionic detergents usually have not, though soap has, but they can have a disinfectant added to them. Hypochlorite, a form of bleach, is mostly so added. It is usually very difficult, or even impossible, to discover what any commercial detergent consists of, and one must rely on the makers' instructions regarding their use.

A *detergent* is a substance which removes dirt and grease by forming a suspension or an emulsion carrying off the dirt in it. The first step is to wet the surface. If water is dropped on top of wax or grease it forms globules, but if a wetting agent is added it forms a fine film. Commercial detergents all have wetting agents in their formula. Usually they include also a water-softening agent that may be something as familiar as washing soda. They may also have a foaming agent added. Foaming is not a necessary part of the cleansing process: it looks effective in the sink, and perhaps on the TV screen, but it can be a nuisance in the washing machine. Soap is a very good anionic detergent in soft water. In hard water it brings out the calcium and magnesium, which cause the hardness, and these form a scum on the water which is difficult to get rid of. The water-softening additive in commercial detergents keeps the calcium and magnesium in solution and avoids the scum: this is their main advantage over soap, but there is nothing magic about them.

The bactericidal, or germ-killing action of hot water is helped by the use of detergents and increased by the use of disinfectants: just as the action of both detergents and disinfectants is increased by heat. A combination of all three is therefore the ideal. In the one-sink method, where dishes are washed by hand, disinfectant is certainly required in addition to detergent. In the two-sink method or in dish-washing machines the temperatures can be so high that detergent only is required, though disinfectant in addition does no harm. It is not possible to give rules about the amount of detergent or disinfectant to add, for the many commercial products differ widely in their formula. It is important to follow the makers' instructions and it is essential to ensure that the required temperature and the correct concentration of chemicals are maintained *throughout* the washing process. Temperatures can fall if machines or sinks are not supervised, and chemicals are used up by contact with grease and food debris, and the solution then needs topping up. Both temperature and concentrations are controlled automatically in modern equipment, but only if the equipment is serviced and inspected regularly. Failure to check washing-up processes, either by hand or machine, can lead to unsatisfactory and perhaps dangerous bacteriological conditions after the wash.

Cloths and brushes. A wet dish-cloth is an unpleasant object. Sodden with suds, swollen with grease and speckled with food particles it is very lacking in aesthetic appeal. There is no corresponding lack of bacterial content, for often such a cloth is teeming with germs. Bacteriologists have taken snips of wet dish-cloths, rubbed them over culture plates and then grown millions of germs from them in the laboratory. These are usually not disease germs, but staphylococci and salmonellae have been grown. In any case, it cannot be considered good hygiene to wipe dishes with wet germ-laden cloths. Nylon brushes are much better as these can be easily cleaned and sterilized.

If dishes have been washed or rinsed in hot water, they will dry in air, and this is the best way to dry them, provided they can be stacked so that water runs *off* them and does not collect in cold pools *on* them. Alternatively they may be dried with sterile cloths. The difficulty is to keep the cloth sterile. Germs will not grow readily on a dry cloth, but the cloth does not remain dry for long, and unless the drying cloth is frequently changed, it too can become charged with germs. Drying sterile dishes with such a cloth undoes the sterilization. If dishes after washing cannot be dried in the air, disposable paper towels can be used in place of cloths. They are virtually germ free, but the difficulty often is to dispose of the disposables.

The Hospital Laboratory

A hospital laboratory is quite a hazardous place wherein to work. There are germs everywhere, many of them very dangerous, but laboratory workers are acutely aware of what they are handling, and on the alert constantly to make sure germs do not spread.

The outlook in a kitchen should be exactly the same. There are germs there too, and they must be kept under control. *There is one big difference.* If anything goes wrong, it is not the kitchen workers who suffer. The results of their mistakes are carried out of the kitchen to the buffet of dining tables. The diner has no defence against a careless chef.

13

The Food Handler

13.1 Introduction

This chapter may be the most boring, as it is full of rules, and rules can be deadening things. Behind the rules of food hygiene there is the life history of countless millions of living unicellular germs striving to establish themselves, often very successfully, in the bodies of multicellular animals and man. If we constantly remain aware of this struggle, the our food hygiene rules cease to be lifeless, but can be seen instead as tactics in a form of bacteriological warfare. Awareness is all.

13.2 Person

The food handler should be a clean person. It appears too obvious to state, yet some cases of food poisoning have been caused simply because the food handler was not clean. The food handler should try not to handle food. When he must, he should do so with hands as clean as possible. How can keep them clean by following certain precautions.

Hand The food handler must keep his or her hands as smooth and soft as possible, and there are many excellent lotions and barrier creams that he can use many times a day, perhaps when going off duty for meals or rest periods. Very hot water, if it contains alkali and detergents, can lead to roughness or sogginess of the skin, especially in the nail folds, and it is wise to wear gloves when one has to immerse the hands for long periods in such water. In general, however, gloves are not advisable, because they can make the hands sweat, and hot, moist hands invite germs to multiply.

Washing must be thorough and frequent. The food handler should instinctively wash his hands before touching food, and always after handling raw meat or vegetables. A dispenser of liquid soap is much more hygienic than a cake of soap, for the latter always has a scum on it, and a pool of turbid water under it, and germs can grow in this. A disinfectant such as hexachlorophane can be added to the liquid soap, and thus help to keep down the number of germs on hands. The dispenser itself must be kept clean, and no clot should be allowed to form on the nozzle. Nail-brushes are best made of nylon brushes for they can be washed and cleaned more easily than hair brushes. They should not be kept immersed in disinfectant fluid, for this often loses its disinfecting properties and becomes contaminated with live germs. The brushes should be changed several times a day. They can be washed in a disinfectant fluid, such as 3 per cent hexachlorophane, and dried thoroughly. Ideally heat-sterilized nail-brushes should be used: they should be available at least in every hospital kitchen. Nail-brushes can remove dirt only if the bristles can reach it, and food handlers ought therefore to keep their nails short. Food-poisoning germs have often been grown from under nails.

Two kinds of germs may be present on the hands, native germs and foreign germs. Native germs normally inhabit and multiply on the skin, often deep down in the pores. Staphylococci are the common native germs. Foreign germs get on the skin accidentally. These may be salmonellae, shigellae or clostridia, in fact any of the food-poisoning germs. There is a very important difference. Foreign germs are not at home on the skin and washing will remove them. Staphylococci are very much at home, and no amount of washing will dislodge them all. In fact scrubbing with hot water may bring the deep staphylococci nearer the surface and make the hands just as dangerous as before. This is one reason why surgeons, after thoroughly 'scrubbing up' in the theatre, still put on sterile gloves before operating. It has even been shown that if a surgeon has a hot shower before operating, he may disperse more staphylococci into the theatre air than he does without a shower. To the food handler the moral is clear. Even though his hands are as clean as he can make them, there may still be staphylococci on them. This is why he should touch food with his hands as little as possible.

Towels, like dish-cloths rapidly become contaminated, and roller towels should never be used. The only satisfactory towel is a paper one, dropped into a foot-operated bin after use. There are many excellent paper towel dispensers on the market, and a good paper towel does dry the hands. Hot-air driers are very hygienic, but they take a long time to get the hands really dry, and busy workers cannot wait.

Hair Staphylococci grow very well on a scurfy scalp, and dandruff can fall over food very easily. This is an unpleasant thought, but food hygienists must face unpleasant facts. Hair must be kept clean and tidy and washed regularly with an anti-dandruff shampoo—there are reliable brands on the market. Caps should be worn, but they should really cover the hair and not be merely frilly ornaments. The *gross bonnet* or 'big hat' of the chef, besides giving him an air of authority, does protect food from dandruff. Hair cream, too, keeps hair in place, and prevents anything falling out of it, and is a good thing if it stops the food handler from throwing his hair back with his hand.

Clothes: Clothes should be light, clean and tidy, and the white apron or coat should be changed at least once daily. Germs cannot jump off clothes on to food, but hands are constantly in contact with clothes and germs can be transferred from one to the other. Soiled clothes in a kitchen, shop or factory are a sign of general slovenliness, and indicate that food hygiene in the workplace is not of a high order.

Health

Food handlers should be aware that illness in themselves can lead to outbreaks of food poisoning. They should therefore report any illness, however mild, and someone with a knowledge of food hygiene should decide whether it is of importance or not. A food handler who has an attack of diarrhoea and vomiting should always be put off duty till cleared by a doctor. Anyone with a nasal discharge, especially if it is purulent, should be examined by a doctor, because germs in a chef's nose can lead to disaster in the dining-room. Sticky eyes may be dangerous for germs are so easily transferred to the hands. Sores, cuts and cracks on the hands are especially important, for often staphylococci get into them and they may be food-poisoning staphylococci. Cuts and sores should always be carefully dressed, and a food handler with a sore should not handle cooked food. If he must, the sore should be covered with a waterproof dressing, but even this has its dangers for it often makes the sore soggy, slow to heal and more attractive to staphylococci. It is, of course, impracticable to put every food handler off duty who cuts his finger, but if the food handler is aware of the danger he can take special care.

Before food handlers are employed they should be questioned about their health, and if they have had any illness of an intestinal type, the doctor may decide to carry out special tests first. These tests, however, are not completely reliable. A person might pass a test one week, and fail it the next, for carriers tend to excrete germs intermittently. The doctor must decide on history as well as on laboratory tests.

Habits

A food handler should have clean habits, and these should be instinctive. He washes his hands before touching food, and after touching food, because he knows that he can pass skin germs on to food, and that he can acquire germs on his hands from raw food, and then transfer it to cooked food. He washes his hands very thoroughly after emptying his bowels, because he knows that germs in the faeces can get through toilet paper on to his hands and so on to food. He is careful about using handkerchiefs, because he knows that though a handkerchief straight from the laundry is sterile, it can be covered with millions of germs after he blows his nose into it. He may, therefore, prefer to use paper handkerchiefs while on duty. He does not smoke on duty, not only because this is forbidden by law, but because he knows it is a dirty habit. Cigarette ash is sterile and will not cause food poisoning if it falls into food, but the food handler can get germs on his hands from his mouth and lips. Moreover, if there is cigarette ash all over the kitchen there are probably all sorts of harmful dust and debris as well. It is a sign of bad hygiene.

The good food handler does not lick his fingers to pick up wrapping-paper, nor does he blow into paper bags to open them. When he slices cold meat, he lifts the slice with a clean knife on to a clean plate or wrapping-paper, not with his bare hands. He uses tongs for lifting cakes and pastries, but, he uses tongs in the back as well as in the front shop. He need not be constantly thinking about food poisoning, but he understands how the germs of food poisoning get around, and he instinctively does everything in his power to make sure he does not help them on their way.

Equipment

The conscientious food handler likes to work with the best equipment, but even if has to make do with something less than the latest, he looks after it well. When he has used a bowl for mixing, he makes sure the bowl is properly washed and sterilized before it is used again. He will clean and sterilize his knives and choppers frequently during the day, and he will make certain that knives used for raw meat are not used for cutting cooked meat. He realizes that table tops can be dangerous pieces of equipment, and he will thoroughly scrub and dry them as often as they become soiled. He will never have raw and cooked food on the same table; if possible not in the same room. He knows that slicing machines, liquidizers, miners and other stainless-steel machines, though they may be bright and gleaming, can easily be contaminated with germs and he will arrange that these machines are dismantled frequently and thoroughly cleaned, if possible at least once a day. The food handler who knows his job takes nothing on trust.

Welfare

The food handler is entitled to good conditions. His working place should be bright, spacious and cheerful, and there should also be good welfare facilities behind it. Toilet facilities should be first-class, with modern water-closets with foot-operated flushes. There should be plenty of changing rooms with lockers for outdoor clothes and footwear. There must be an adequate number of wash-basins, properly equipped with soap dispensers and paper towels, and there should also be a supply of tissue-paper handkerchiefs. A rest room for off-duty periods should be clean and comfortable and canteen should set the highest standard of hygiene. All this may sound very ideal and impracticable in the small shop, kitchen or factory, but only the ideal is good enough in food production. If the small establishment cannot reach an acceptable standard, it should be eliminated. Too many outbreaks have been caused by ignorance and filth.

Knowledge

Equipment, welfare, habits, health and hands — they are all important — but knowledge is more important still. Everyone who handles food should have some knowledge of food poisoning. A little learning may be a dangerous thing, if it is not digested and understood. Knowledge, on the other hand, implies an understanding of facts. The food hygienist and the food handler need all the knowledge they can acquire.

